Dear Manfred,

I hope my book will electrify you

Hans
July 30, 2001

100 YEARS OF POWER PLANT DEVELOPMENT

Thomas Alva Edison's Visit to the Power Plant Moabit in Berlin, Germany, With Emil Rathenau on September 25, 1911[3]

100 YEARS OF POWER PLANT DEVELOPMENT

Focus on Steam and Gas Turbines as Prime Movers

Power Plant Cycles and Concepts for
Fossil Steam Turbine Power Plants
Nuclear Power Plants
Co-Generation Plants
Gas Turbine Power Plants
Combined-Cycle Power Plants
Repowering Steam Turbine Plants With Gas Turbines
Coal Gasification and Fuel Cell Combined-Cycle Plants

BY

Heinz Termuehlen

ASME PRESS   NEW YORK   2001

Copyright © 2001 by
The American Society of Mechanical Engineers
Three Park Avenue, New York, N.Y. 10016

All rights reserved. Printed in the United States of America. Except as permitted under the United States Copyright Act of 1976, no part of this publication may be reproduced or distributed in any form or by any means, or stored in a database or retrieval system, without the prior written permission of the publisher.

Statement from By-Laws: The Society shall not be responsible for statements or opinions advanced in papers . . . or printed in its publications (7.1.3)

INFORMATION CONTAINED IN THIS WORK HAS BEEN OBTAINED BY THE AMERICAN SOCIETY OF MECHANICAL ENGINEERS FROM SOURCES BELIEVED TO BE RELIABLE. HOWEVER, NEITHER ASME NOR ITS AUTHORS OR EDITORS GUARANTEE THE ACCURACY OR COMPLETENESS OF ANY INFORMATION PUBLISHED IN THIS WORK. NEITHER ASME NOR ITS AUTHORS AND EDITORS SHALL BE RESPONSIBLE FOR ANY ERRORS, OMISSIONS, OR DAMAGES ARISING OUT OF THE USE OF THIS INFORMATION. THE WORK IS PUBLISHED WITH THE UNDERSTANDING THAT ASME AND ITS AUTHORS AND EDITORS ARE SUPPLYING INFORMATION BUT ARE NOT ATTEMPTING TO RENDER ENGINEERING OR OTHER PROFESSIONAL SERVICES. IF SUCH ENGINEERING OR PROFESSIONAL SERVICES ARE REQUIRED, THE ASSISTANCE OF AN APPROPRIATE PROFESSIONAL SHOULD BE SOUGHT.

For authorization to photocopy material for internal or personal use under circumstances not falling within the fair use provisions of the Copyright Act, contact the Copyright Clearance Center (CCC), 222 Rosewood Drive, Danvers, MA 01923, Tel: 978-750-8400, www.copyright.com.

Requests for special permission or bulk reproduction should be addressed to the ASME Technical Publishing Department.

Library of Congress Cataloging-in-Publication Data
Termuehlen, Heinz, 1936 –
100 years of power plant development: focus on steam and gas turbines as prime movers/by Heinz Termuehlen.
   p. cm.
  Includes bibliographical references and index.
  ISBN 0-7918-0159-4
1. Electric power-plants — History.
2. Steam turbines — History.
3. Gas-turbines — History.
I. Title: Hundred years of power plant development.
TK1001.T45 2001
   2001024970

PREFACE

I have been involved in the power plant developing process in Europe since 1958 and since 1969 in the United States. After working in the power generation industry for more than 40 years, I retired in August of 1999. Not long after that I received a phone call from Mary Grace Stefanchik of ASME Press inviting me to write a book based on my extensive experience and the almost 100 technical papers I have written on a wide variety of subjects concerning power generation issues.

I was ready for a new challenge. I accepted the invitation and proposed to write this book, *100 Years of Power Plant Development: Focus on Steam and Gas Turbines as Prime Movers*. My broad experience in the power generation industry has afforded me knowledge of all types of power plant concepts of the past and present, and with that I present my forecast for the future of power generation, applying the most advanced technologies.

Heinz Termuehlen

ACKNOWLEDGMENTS

Reference to illustrations and/or information utilized in this book is given in the text in the form of numbered references. The following societies and companies have provided permission to publish illustrations and/or information from 81 books and papers covering various topics of power generation:

- American Nuclear Society (ANS)
- American Society of Mechanical Engineers (ASME)
- Berliner Kraft- and Licht Aktiengesellschaft (BEWAG)
- Brennstoff-Waerme-Kraft (VDI-Verlag)
- Canadian Gas Association (IAGT)
- Electric Power Research Institute (EPRI)
- GE Marine & Industrial Engines
- GE Power Generation
- Institution of Mechanical Engineers (IMechE)
- Rocky Mountain Electrical League
- PennWell Publishing Company [Power-Gen and American Power Conference (APC)]
- Siemens Power Generation Group
- Springer Verlag
- Vericor Power Systems
- VGB Kraftwerkstechnik GmbH

The author is thankful for the permission provided by these organizations, which has made it possible for him to give a historic overview of the evolutionary power plant development. The large number of illustrations help the reader follow this progress. As engineers like to say: "A good drawing is better than a thousand words."

CONTENTS

Preface		v
Acknowledgments		vii

Introduction			1
Chapter 1	Historical Review of Power Generation With Steam and Gas Turbines		3
Chapter 2	Fossil Steam Turbine Power Plants		9
Chapter 3	Nuclear Power Plants		51
Chapter 4	Co-Generation Plants		65
Chapter 5	Gas Turbine Power Plants		83
Chapter 6	Combined-Cycle Power Plants		111
Chapter 7	Repowering Steam Turbine Plants With Gas Turbines		131
Chapter 8	Coal Gasification and Fuel Cell Combined-Cycle Plants		155
Chapter 9	Future Power Generation		171

References	189
About the Author	197
Index	199

INTRODUCTION

In the year 2000 80% of the world's power was produced by steam or gas turbine prime movers generating about 15×10^9 MWh annually.[1] A large variety of power plant concepts have been developed to best utilize steam and/or gas turbines for each specific application. All power plant concepts have been developed over many decades in an evolutionary mode beginning roughly 100 years ago with steam turbine power plants.

The power generation industry became one of the most important global industries generating electricity for all other industries as well as for commercial and domestic use. Thomas Edison realized the global importance of electric lighting and power generation. In 1883, the "Deutsche Edison-Gesellschaft" (German Edison Company) was founded by Thomas Edison and Emil Rathenau.[2] Werner von Siemens, who designed and built the first dynamo for power generation in 1866, received the license to manufacture Edison light bulbs from this company. The frontispiece of this book shows Thomas Edison with Emil Rathenau in the power station Moabit in Berlin, Germany.[3]

The global power generation industry has the mission to generate for all users uninterrupted electric power under all kinds of weather conditions at any level of demand, from peaking demand at daytime to extremely low demand at night. Electric power has to be generated at the most economic, low-cost level. The power generation industry depends on the available fuel sources and has to develop and apply the best technologies for converting the fuel into electric power with the lowest adverse effect on the environment.

For the power generation industry to achieve its goal, power plants have to be developed, built, and operated based on the most advanced technologies and practices. This is an extremely challenging task for engineers in the power generation industry. In addition, the availability of fuels and fuel utilization can become not only a technical and economical issue, but also a political one, e.g., the political issues related to the use of nuclear fuel and the emission reduction when burning fossil fuels.

The utilized fuels influence steam and gas turbine designs. For example, nuclear reactors provide very different steam conditions than fossil boilers for steam turbines, and burning coal-derived gas requires major changes in gas turbine combustion systems.

This book describes the evolution of the large variety of power plant concepts for any kind of fuel. It shows how advances in power plant technologies, especially the large steam and gas turbine design, have improved the performance of power stations. These advanced technologies can also be applied to existing power plants, for example, in repowering projects. The evolution of large steam turbine and heavy-duty gas turbine designs toward advanced steam and gas turbines often required to find solutions to operational problems. This book shows how problems were overcome and how this experience can help to solve problems in the future.

Advances have been made to build not only more efficient power plants, but also much more environmentally friendly power plants. The environmental effect of power plants is greatly influenced by the fuel burned, and the rightful comparison of the emissions of power plants burning different fuels is difficult. Besides the pollutants of sulfur dioxide and nitrogen oxides, the discharge of carbon dioxide of different power plant concepts is also addressed.

CHAPTER 1

Historical Review of Power Generation With Steam and Gas Turbines

At the turn of the century, the industrial revolution prompted the development of steam turbines to generate electric power. Until that time, steam engines with increasingly larger flywheels functioning as generator rotors were used. The first electric power was generated by a steam turbine-generator in 1884. The rising demand for electric power led to an increase of unit ratings and warranted the quick development of larger steam turbine-generators. The 1906 photograph of the machine house in Figure 1-1 reveals the size difference between a 3000 kW steam engine (in the foreground) and the two smaller 3000 kW and one 6000 kW steam turbine generators. In their early applications, steam turbines generated electric power in coal-fired power plants. There was a steady improvement of power plant performance achieved by increasing steam conditions, cycle optimization, and increasing equipment efficiencies. More and more efficient steam turbines were also used as mechanical drives for industry use, marine propulsion, and as industrial prime movers for co-generation of electrical power and process steam or heat.

In the 1940s and early 1950s nuclear research reactors were put into operation. Some of these plants already generated electric power. However, the first reactors for commercial power generation were built in the mid-1950s. After that, larger boiling water reactors and pressurized water reactors were specifically designed for nuclear power plants with saturated main steam driven large half-speed turbines.

Starting in the 1930s to 1940s gas turbines were developed and put into operation as aircraft engines. With some earlier exceptions, the significant utilization of gas turbines for power generation began in the 1960s when peaking power plants and co-generation plants were built.

In the 1960s combined-cycle power plants were built as the most efficient power plant design concept. At the same time, repowering of existent steam plants with gas turbines was promoted as a way to improve the performance of steam plants and to increase their capacity.

All of these power plant concepts have their specific advantages and have proven operating experience. Advances in power plant concepts have been made by increasing unit size and the overall plant efficiency. Besides power plant efficiency, there are equally important issues such as fuel availability and fuel flexibility as well as emissions and operating flexibility. When evaluating the application of plant concepts, all these issues have to be addressed.

6 ● **100 YEARS OF POWER PLANT DEVELOPMENT**

FIGURE 1-1. Power plant with steam engines and steam turbines.

In general, all power plant concepts can be modified to not only generate electric power but also supply steam or heat for industrial use or district heating needs, which are then called co-generation plants. Co-generation plants are desired because the fuel utilization to produce electric power and useful thermal energy is very high.

This book describes the different power plant concepts, their application, and performance. It also discusses the history of power plant concepts, gives examples and discusses their general utilization, and shows even more advanced plant concepts for future consideration. The historical development of power plants is best illustrated in Figure 1-2, which shows their unit rating growth over the last century.

- Fossil-fueled steam turbine power plants are by far the major power generation technology applied. Their development process started around 1900. About 30 years later power plants with unit ratings above 100 MW were built and in another 30 years the unit size reached 1000 MW. The maximum unit size of 1300 MW

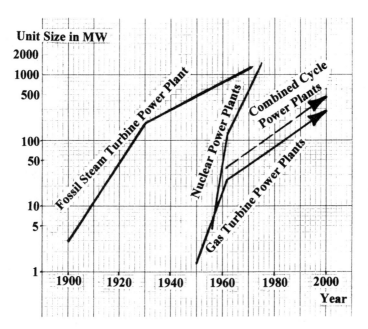

FIGURE 1-2. Historical development of power plants.

was reached in the early 1970s utilizing a cross-compound (two separate turbine-generator shafts) steam turbine design. Since then unit rating has not been increased. Presently, the maximum unit rating considered is about 1000 MW, utilizing a tandem-compound (one turbine-generator shaft) steam turbine design. Even though the unit size of steam turbine plants has not increased, advances have been made in the last decade to build much more efficient fossil-fueled power plants with a major reduction in their emissions.

- Nuclear power plants had a very short evolution process. Within 20 years the unit rating of 1300 MW was reached and became a standard size. The technology was available to even increase the unit rating to a 2000 MW level; however, this step was not taken and only a few nuclear power plant units are rated at about 1500 MW. Presently, there is very little activity in the building of nuclear power plants for mainly political and economical considerations.

- Gas turbines were developed as aircraft engines in the 1930s and 1940s. Because of their operating flexibility they were first utilized for power generation to provide peaking power. Later, heavy-duty gas turbines were introduced for power generation. The unit size of gas turbines increased in roughly 30 years to 100 MW for 60 Hz applications. Today, gas turbines up to 250 MW unit size are available. It is projected that the gas turbine unit size will continue to increase; however, the history of both fossil and nuclear power stations show that the unit size levels off. The reason for such a rating plateau could be due to technological, political, or economical considerations.
- Combined-cycle power plants with gas and steam turbines can be built in a large variety of plant concepts. If the exhaust energy of one gas turbine is utilized to generate steam for one steam turbine, the combined-cycle plant size is about 50% larger than a peaking plant with a gas turbine alone. Such combined-cycle 60 Hz power plants are available today with unit ratings of roughly 370 MW. The power plant output can be doubled by building a combined-cycle plant with two gas turbines and one steam turbine, which then generates a total of 740 MW. Heavy-duty gas turbines for 50 Hz applications are even larger because of the scaling technique that can be applied, reaching for a two-gas turbine and one steam turbine arrangement already the 1000 MW threshold.

CHAPTER 2

Fossil Steam Turbine Power Plants

Sir Charles Parsons built the first steam turbine-generator, referred to as No.1, in 1884. This unit, illustrated in Figure 2-1, was rated at 7.5 kW and featured an axial double-flow steam turbine with reaction-type blading directly coupled to its generator. Because of its low thermal efficiency of only 1.6%, it was called a "steam eater."[4]

Improvements in size and efficiency followed immediately and by 1889 Parsons had formed the Newcastle and District Electric Lighting Company (DISCO), which operated the first central power station with alternating-current generators. This station, named Forth Banks in Newcastle, was equipped with two 75 kW units attaining an efficiency of 6%. As shown in the diagram of the same figure, a thermal efficiency of 8.3% was reached with the first condensing turbine rated 100 kW in 1891, and a thermal efficiency of 21% was achieved by a 5 MW steam turbine-generator in 1903.

Power generation stations with steam turbine-generators as prime movers replaced the original power plants with steam engines. These early power plants featured a boiler, a turbine, a generator, a condenser, and a boiler feed pump forming a thermal cycle as illustrated in Figure 2-2. The feedwater pump supplies pressurized feedwater to the boiler, which features an economizer, evaporator, and a super heater. The economizer heats the feedwater before it is boiled in the evaporator and then superheated in the boiler's superheater section. Expansion of this superheated main steam to the condenser pressure level takes place in the turbine, which drives the generator to produce electric power. The exhaust steam of the turbine is condensed in the condenser with low temperature cooling water establishing vacuum conditions. Draining the condensate to the boiler feedwater pump (condensate pump) is the last step in closing the steam/water cycle.

In early power plants, built around 1900, the turbine-generators were rated in a range from 1 MW to 10 MW. Two plant arrangements were utilized, namely a vertical and a horizontal, as shown in Figure 2-3.[5] As turbine unit sizes became larger, especially when multiple casing turbines were introduced, only the horizontal turbine arrangement with a directly coupled generator and the condenser arranged underneath the turbine exhaust section were applied.

Early power plants were operated with low main steam conditions in the pressure range from 10 bar to 20 bar (150 psia to 300 psia) and in

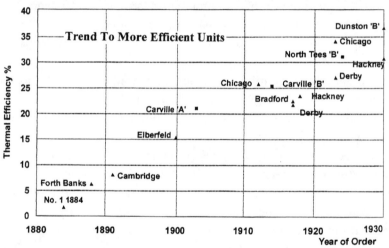

FIGURE 2-1. First turbine-generator built in 1884 and the trend to more efficient units.

the temperature range from saturated steam to slightly superheated steam [50 K to 100 K (90°F to 180°F)]. They did not feature any feedwater or regenerative heating and no reheat. The overall efficiency of these early power stations from the coal combustion to the net plant electric output was roughly 15%.[5]

It is important to realize that the total efficiency of a power plant is greatly affected by its thermal cycle. Even the ideal Carnot cycle efficiency of the early power plants, as illustrated in Figure 2-4, was very low because of the low main steam temperature level. The Carnot cycle

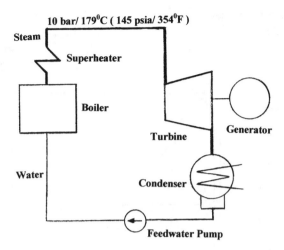

FIGURE 2-2. Early steam turbine power plant concept.

as the ideal thermal cycle cannot be realized in a power plant; however, it shows the ideal thermal cycle efficiency for a given temperature difference.

The Carnot cycle forms a rectangular field in the temperature-entropy diagram with an isentropic compression from A to B, an isothermal ex-

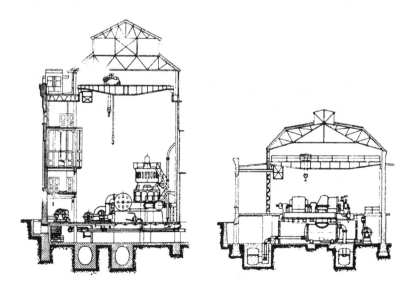

FIGURE 2-3. Early steam turbine machine house arrangements.

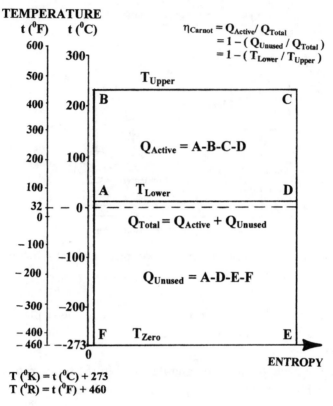

FIGURE 2-4. Carnot process of early steam turbine power plant.

pansion from B to C, an isentropic expansion from C to D, and an isothermal compression from D to A, closing the cycle. The Carnot thermal cycle efficiency is calculated by dividing the active thermal energy

$$Q_{active} \triangleq \text{(Rectangular A-B-C-D)}$$

by the total energy

$$Q_{total} \triangleq \text{(Rectangular B-C-E-F)}$$

Since the Carnot cycle forms rectangles with isentropic and isothermal changes of state in the temperature-entropy diagram, the following equation applies:

$$\eta_{Carnot} = Q_{active}/Q_{total}$$

$$\eta_{carnot} = 1 - (Q_{unused}/Q_{total}) = 1 - (T_{lower}/T_{upper})$$

For this ideal process the thermal cycle efficiency is only a function of two temperatures that must be applied as absolute temperature values in either Kelvin (K) or Rankine (°R):

$$T(K) = t(°C) + 273.15 \quad \text{or} \quad T(°R) = t(°F) + 459.67$$

A Carnot cycle efficiency comparison of an early steam turbine power plant with two modern plants should show the cycle efficiency improvements. For all three plants the same condensation temperature of $t = 10°C$ (50°F) has been assumed:

$$T_{lower} = 10°C + 273 = 283 \text{ K} \quad \text{or} \quad T_{lower} = 50°F + 460 = 510°R$$

The main steam conditions for an early plant with 10 bar (145 psia) pressure and 50 K (90°F) superheating provides an upper temperature of:

$$t_{upper} = t_{saturated} + 50 = 179 + 50 = 229°C$$

$$T_{upper} = t_{upper} + 273 = 502 \text{ K}$$

or

$$t_{upper} = t_{saturated} + 90 = 354 + 90 = 444°F$$

$$T_{upper} = t_{upper} + 460 = 904°R$$

Please note that the amount of superheating is a temperature difference and therefore 1°F = 5/9 K. For temperature differences, the metric system uses degree Kelvin (K) instead of degree Celsius (°C).

With these temperatures the Carnot cycle efficiency for the early steam turbine power plant can be calculated as follows:

$$\eta_{Carnot} = 1 - (T_{lower}/T_{upper}) = 1 - (283/502) = 1 - 510/904 = 43.6\%$$

The steam temperatures of modern steam turbine power plants are much higher. Assuming a steam temperature of 565°C (1050°F) provides a Carnot cycle efficiency of:

$$T_{upper} = 565 + 273 = 838 \text{ K}$$

$$\eta_{Carnot} = 1 - (283/838) = 66.2\% \text{ (modern steam turbine power plant)}$$

These results reveal that the ideal thermal cycle efficiency of steam turbine power plants has been drastically improved. An even further improvement has been achieved with modern combined-cycle power plants having an upper temperature of 1300°C (2372°F):

$$T_{upper} = 1300 + 273 = 1573 \text{ K}$$

$$\eta_{carnot} = 1 - (283/1573) = 82\% \text{ (combined-cycle power plant)}$$

This combined-cycle power plant Carnot efficiency is about twice as high as the efficiency of the early steam turbine power plants. However, it must be realized that the Carnot efficiencies represent the theoretical upper performance for ideal thermal cycles, assuming that heat is provided isothermally at the upper temperature. Correcting just with an average temperature at which the heat is actually provided in the combustion chamber of a gas turbine at constant pressure reduces the 82% to roughly 73%.

Returning to the example of an early steam turbine power plant, there is quite a difference between the stated total power plant efficiency of about 15% and the calculated 43.6% Carnot cycle efficiency. The actual thermal cycle of a steam turbine power plant is much better represented by the Rankine steam cycle, as illustrated in Figure 2-5. The Rankine process, also referred to as the Clausius-Rankine process, of the early steam turbine power plant begins with the heating of water in the boiler (from A to B). The water is then boiled to become saturated steam (from B to C') at a pressure of 10 bar (145 psia). The steam is superheated by 50 K (90°F) to become main steam for the steam turbine (from C' to C). Expansion of the steam takes place in the steam turbine (from C to D).

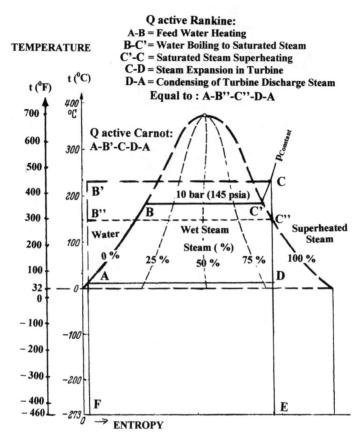

FIGURE 2-5. Rankine process of early steam turbine power plant.

The turbine exhaust steam is condensed in the condenser, closing the cycle (from D to A). This Rankine process of the early steam turbine power plant provides active heat energy Q_{active} within the area A-B-C-C'-D-A, which covers about two-thirds of the rectangular area of the Carnot process (A-B'-C-D-A).

The Rankine cycle representing the steam turbine power plant cycle provides the tool to calculate the plant's actual thermal cycle performance. The active Rankine cycle area of the shown example can be expressed by an average of two-thirds of the active Carnot cyle temperature because both processes have the same lower temperature and cover the same entropy range, which is the area A-B"-C"-D-A:

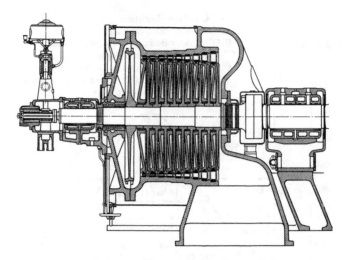

FIGURE 2-6. Early steam turbine design.

$$T_{\text{active Carnot}} = T_{\text{upper Carnot}} - T_{\text{lower}} = 502 - 283 = 219 \text{ K}$$

$$T_{\text{active Rankine}} = T_{\text{active Carnot}} \times 2/3 = 219 \times 2/3 = 146 \text{ K}$$

$$T_{\text{upper Rankine}} = T_{\text{lower}} + T_{\text{active Rankine}} = 283 + 146 = 429 \text{ K}$$

$$\eta_{\text{Rankine}} = 1 - (T_{\text{lower}} / T_{\text{upper Rankine}}) = 1 - (283/429) = 34.0\%$$

Improvement of this early power station thermal cycle efficiency has been realized not only by raising the steam conditions, but also by preheating the feedwater with feedwater heaters and by reheating the steam in the boiler after it was partially expanded in a turbine high pressure section.

The calculated 34.0% Rankine cycle efficiency still does not represent the total power plant performance for the early steam turbine power plant of roughly 15%. Both the Carnot and Rankine cycles assumed an isentropic or adiabatic expansion in the steam turbine (from C to D) that represents a turbine thermodynamic efficiency of 100%. However, early steam turbines had a thermodynamic efficiency of only about 60%. A typical example of such early single-casing steam turbine design is depicted in Figure 2-6, featuring a two-stage, partial-arc admission and

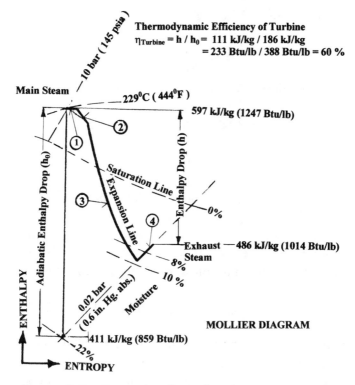

FIGURE 2-7. Expansion line of early steam turbine.

nine impulse-type stages in a horizontal arrangement with a downward exhaust.[5]

The thermodynamic efficiency of a steam turbine is represented as an expansion line in an enthalpy-entropy diagram commonly referred to as the Mollier diagram. The expansion line of the early steam turbine is illustrated in Figure 2-7 and shows the following four sections:

1. pressure drop upstream of the control stage
2. expansion in the partial-arc control stage
3. expansion in the full-arc blade group
4. exhaust losses downstream of the last-stage blade row

The expansion line of an early steam turbine is depicted in the enthalpy-entropy diagram with the enthalpy drop of h. The thermodynamic efficiency of the steam turbine is calculated by dividing its enthalpy drop through the adiabatic enthalpy drop h_0:

Adiabatic expansion:

$$h_0 = 1247 - 859 = 388 \text{ Btu/lb}$$

Turbine enthalpy drop:

$$h = 1247 - 1014 = 233 \text{ Btu/lb}$$

Thermodynamic turbine efficiency:

$$\eta_{Turbine} = h/h_0 = 233/388 = 60\%$$

With the steam cycle efficiency of 34.0% and a thermodynamic turbine efficiency of 60%, the thermal efficiency of a steam turbine power plant can be calculated by the multiplication of these two values.

$$\eta_{Thermal} = \eta_{Rankine} \times \eta_{Turbine} = 0.34 \times 0.60 = 20.4\%$$

Since this plant thermal efficiency still does not include the mechanical turbine-generator and auxiliary losses nor the generator and boiler efficiencies, the total steam turbine power plant efficiency can be estimated by multiplication of the following efficiencies:

Generator efficiency (including mechanical losses):

$$\eta_{Generator} = 91\%$$

Boiler efficiency (including pipe system losses):

$$\eta_{Boiler} = 83\%$$

Auxiliary efficiency (including boiler feed pump):

$$\eta_{Aux} = 97\%$$

Total steam turbine power plant efficiency:

$$\eta_{Power\ plant} = \eta_{Thermal} \times \eta_{Generator} \times \eta_{Boiler} \times \eta_{Aux}$$

$$\eta_{Power\ plant} = 20.4\% \times 91\% \times 83\% \times 97\% = 15\%$$

This steam turbine power plant performance has been improved mainly by raising the main steam conditions and improving the efficiencies of power plant components, especially the thermodynamic efficiency of the steam turbine.

Early steam turbines operated mostly below the saturation line. However, the turbine exhaust steam had only a relatively low moisture content, less than 10%. A low moisture content of steam in the last-stage turbine blade section is very important, since experience has shown severe water droplet erosion on last-stage moving blades when operating in a high moisture content environment. The maximum limit is typically given at 13%; however, the lowest possible moisture content should always be considered a power plant cycle layout goal.

Figure 2-8 shows how the thermodynamic turbine efficiency and changing steam conditions influence the moisture content in the steam turbine. The five expansion lines (ELs) illustrate the following effects:

- EL 2: raising steam turbine efficiency to 80%
- EL 3: raising main steam pressure to 28 bar (400 psia)
- EL 4: raising main steam temperature to 399°C (750°F)
- EL 5: raising main steam pressure to 83 bar (1200 psia) and main steam temperature to 538°C (1000°F)
- EL 5a: raising condenser pressure to 0.068 bar (2 in. Hg. abs)

By increasing the efficiency of the early steam turbine of 60% (EL 1) to a thermodynamic turbine efficiency of 80% (EL 2) without changing the steam conditions, a drastic increase in moisture content must be expected. As shown in this figure, the turbine expansion line endpoint without exhaust loss increased from less than 10% to about 17% moisture at an unchanged back pressure of 0.02 bar (0.6 in. Hg. abs.). Raising the main steam pressure leads to an even higher moisture content, as illustrated by the expansion line (EL 3), for a main steam pressure of 28 bar (400 psia), the moisture content grows to 25%. On the other hand raising the main steam temperature without change of the main steam pressure would reduce the moisture level to less than 8%, as illustrated by expansion line (EL 4).

However, in the early days of steam turbine power plant development, high-temperature materials for turbine and boiler components as well as main steam piping were not yet available. Later steam turbine and boiler design improvements and the introduction of alloy steels for high-temperature components allowed main steam conditions to rise to 83 bar/

22 ● **100 YEARS OF POWER PLANT DEVELOPMENT**

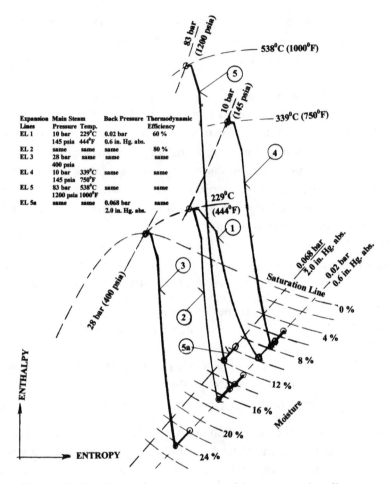

FIGURE 2-8. Non-reheat steam turbine expansion lines.

538°C (1200 psia/1000°F). The expansion line (EL 5) shows that for these conditions, the moisture content is about 15% at 0.02 bar (0.6 in. Hg. abs.) condenser pressure. This moisture content is still too high for long-term operation without droplet erosion damage on the last-stage blades. Drain systems in the low-pressure turbine stages have been developed to reduce the moisture content at the last-stage blade. More importantly, the backpressure of steam turbines increased, especially for power plants with cooling towers. Steam turbines were mostly designed for back-pressures of 0.068 bar to 0.12 bar (2.0 in. Hg. abs. to 3.5 in. Hg. abs.). The expansion line (EL 5a) reveals that the back pressure of a design with 0.068 bar (2.0 in. Hg. abs.) leads to a moisture content of less than

Fossil Steam Turbine Power Plants • 23

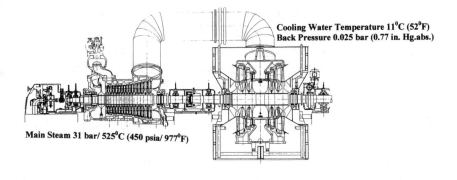

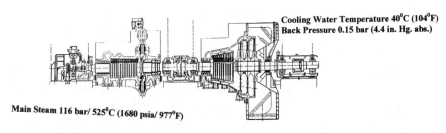

FIGURE 2-9. Two 60 MW non-reheat steam turbine designs for different cooling water temperatures.

12%, which is still high but acceptable for a well-designed last-stage blade with leading edge protection such as flame-hardening or stelite strips.

In Figure 2-9 two turbine designs reveal how important the optimal main steam pressure depends for non-reheat turbines on the turbine exhaust conditions.[6] Both 60 MW turbines are of a two-casing design, however the first unit operates with fresh cooling water of 11°C (52°F), while the second plant features a cooling tower design for a cooling water temperature of 40°C (104°F). The different cooling water conditions provide the first turbine with a cooling water-to-condensate ratio of 75, a condenser pressure of 0.025 bar (0.74 in. Hg. abs.), and the second turbine with a cooling water-to-condensate ratio of 45, a condenser pressure of 0.15 bar (4.4 in. Hg. abs.). For both turbines a main steam temperature of 525°C (977°F) was selected. However, the main steam pressure of the first turbine was as low as 31 bar (450 psia) to avoid a too high last-stage moisture content, whereas the second turbine's main steam pressure could be raised to 116 bar (1680 psia).

Both turbines feature a single-flow HP turbine, whereas the LP turbine section of the first unit is a double-flow design for the large volumetric flow at the low backpressure. The volumetric flow (specific volume) of the second unit is about 40 times smaller and, therefore, only a single-flow LP turbine section with even a smaller last-stage blade length is needed. The HP turbine is turned around to balance the thrust force of the HP turbine and single-flow LP turbine. This example also reveals that the turbine size is more a function of the volumetric flow rather than the output, and therefore the cost of steam turbines can hardly be estimated in cost per kilowatt output. A better measure is, for example, the LP turbine annular area in cost per square foot.

With the introduction of power plants with reheat steam turbines, it was possible to raise the main steam pressure without the concern of a too high moisture content at the LP turbine's last-stage blades. The expansion line of a reheat steam turbine is illustrated in Figure 2-10 with main steam conditions of 166 bar/538°C (2400 psia/1000°F). The main steam is expanded in the HP turbine section to the reheat steam pressure of 45 bar (650 psia) and then reheated in the boiler's reheater section to a reheat steam temperature of 538°C (1000°F). Pressure losses of 4 bar (55 psia) in the reheating system reduce the intermediate pressure (IP) turbine inlet pressure to 41 bar (595 psia). In the IP turbine the reheated steam expands to LP turbine inlet conditions of 7 bar/ 288°C (100 psia/ 550°F). Because of the low IP turbine inlet pressure the IP/LP turbine expansion line ends at a very low moisture content of 10% with a backpressure of 0.068 bar (2.0 in. Hg. abs.), despite the high efficiencies reflected in the steep expansion lines of all the turbine sections. Also indicated is the effect of an internal turbine second last (L-1) stage drainage at a moisture level of 6%, which reduces the turbine end moisture by roughly 0.5%.

The thermal cycle of a reheat steam turbine power plant is again best revealed in a temperature-entropy diagram. As shown in Figure 2-11, the Rankine cycle of a single reheat turbine with feedwater preheating to a high final feedwater temperature provides the following cycle performance improvements. Compared to the early steam turbine non-reheat cycle without feedwater preheating and a Rankine steam cycle efficiency of only 34%, the shown reheat cycle provides a Rankine cycle efficiency of 55.6%. The large improvement has been achieved by not only raising the steam conditions, but by adding single reheat to the cycle "(+) area" and by preheating the feedwater with feedwater heaters being supplied with extraction steam from the turbine. This has been shown as a "(−) area" because the heat for feedwater heating does not have to be pro-

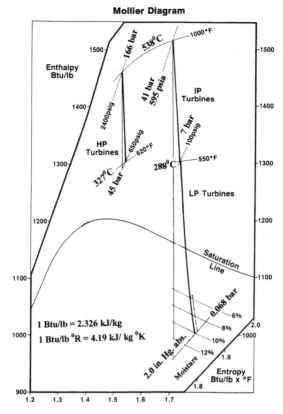

FIGURE 2-10. Expansion line of reheat steam turbine.

vided by the boiler. The addition of the high energy level reheat and the deletion of the low energy level feedwater preheating provide significant cycle improvement.[7] The cycle efficiency advantage only provides a major power plant performance advantage if the low energy level (low temperature level) heat can efficiently be used in the boiler. This became reality with the introduction of pulverized coal combustion that can utilize combustion air with a temperature of up to at least 350°C (660°F). Preheating of combustion air with low temperature flue gas in regenerative air heaters replaced the preheating of low temperature condensate in the boiler, keeping the boiler efficiency level up.

With the Rankine cycle efficiency of 55.6% and thermodynamic turbine efficiency of 82% the early reheat steam turbine power plants achieved a total power plant efficiency of:

26 ● 100 YEARS OF POWER PLANT DEVELOPMENT

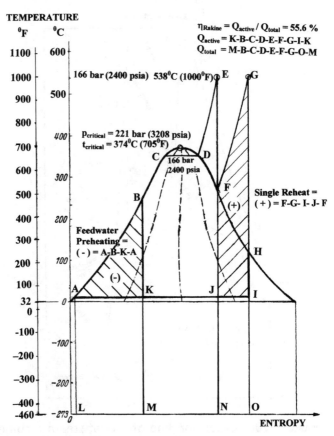

FIGURE 2-11. Single reheat steam cycle with feedwater preheating.

$$\eta_{Thermal} = \eta_{Rankine} \times \eta_{Turbine} = 55.6\% \times 82\% = 44.6\%$$

$$\eta_{Generator} = 97\% \text{ (including mechanical losses)}$$

$$\eta_{Boiler} = 88\% \text{ (including pipe system losses)}$$

$$\eta_{Aux} = 95\% \text{ (including boiler feed pump)}$$

$$\eta_{Power\ plant} = \eta_{Thermal} \times \eta_{Generator} \times \eta_{Boiler} \times \eta_{Aux}$$

$$\eta_{Power\ plant} = 44.6\% \times 97\% \times 88\% \times 95\% = 36.2\%$$

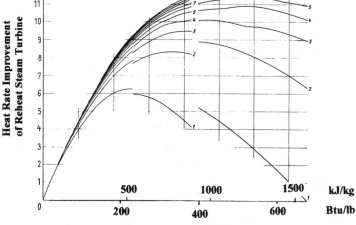

FIGURE 2-12. Power plant cycle improvement as a function of feedwater preheat enthalpy gain and number of feedwater heaters.

This is quite an improvement over the original plant efficiency of 15%. The introduction of feedwater heating and reheat alone had a major effect in improving power plant efficiency. Modern steam turbine power plants are equipped with seven to ten feedwater heaters utilizing steam turbine extraction steam to preheat the boiler feedwater. In Figure 2-12 the power plant efficiency improvement by feedwater preheating is plotted over the feedwater enthalpy gain for a typical single reheat cycle with 180 bar/535°C/535°C (2600 psia/995°F/995°F) main and reheat steam conditions.[6] The application of feedwater preheating can account for up to 13% power plant efficiency improvement. The addition of reheat to the cycle results in an improvement in steam turbine power plant efficiency by roughly 5% for single reheat and an additional 2.5% for double reheat. However, the actual improvement for feedwater preheating and reheat depends heavily on the power plant overall cycle conditions. Only a high main steam pressure above supercritical pressure of 221 bar (3208 psia) justifies the application of double reheat and ten feedwater heaters with a high final feedwater temperature. For such a steam pressure level, once-through boilers that were also applied to subcritical steam conditions in Europe, replaced the drum-type boilers.

The trend to more efficient steam turbine power plants led in the 1950s to a variety of unique HP turbine designs for very high steam conditions. Figure 2-13 shows three examples, starting in 1951 with a 600°C (1112°F) main steam temperature unit and leading in 1960 to a 355 bar/650°C (5000 psia/1200°F) unit.[8,9,10] All three HP turbine designs feature multiple shell casings with the inlet sections made from austenitic high-alloy steel. The first two turbines utilize barrel-type outer casings with radial joints, while the third turbine model is of a horizontal joint design. The first HP turbine shown is of a radial flow path design. Operating experience with such high main steam conditions proved that this technology provides very reliable and efficient power plant performance. However, austenitic forgings and castings for turbine and boiler components as well as for steam piping were about seven times more expensive than the low-alloy steel forgings and castings generally applied.

The two diagrams in Figure 2-14 illustrate the evolution of main steam pressure and temperature levels up to the main steam conditions of the high performance HP turbines built in the 1950s. It can be seen that at this time there was quite a dynamic development leading to supercritical pressures. This development was possible due to the introduction of once-through steam generators. The introduction of reheat steam turbine power plants in the 1920s to 1930s allowed the first step change in the diagram to pressure levels above 120 bar (1740 psia). The temperature trend did not experience such step changes; it shows a typical development curve with an almost constant slope and a smaller pitch at the end. However, increasing the steam temperature by more than 100%, required major developmental effort in design and metallurgy for steam turbines, steam generators, and steam piping.

Despite the availability of high steam temperature technology, most of the power plants in the 1950s through the 1980s were built for only moderate steam conditions. Their main steam pressure ranged from 124 bar to 241 bar (1800 psia to 3500 psia) and their main steam temperature ranged from 510°C to 566°C (950°F to 1050°F). This indicates that the most advanced technology is only utilized if it can also be economically

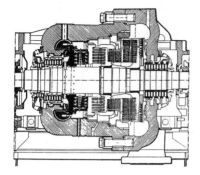

Leverkusen #2 1951
146 bar/600°C (2130 psia/ 1112°F)

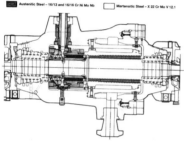

Huels 1956
292 bar/ 600°C (4240 psia/ 1112°F)
Leverkusen #5 1957
146 bar/ 640°C (2130 psia/1184°F)

Eddystone 1960
345 bar/649°C(5000 psia/1200°F)

FIGURE 2-13. Advanced HP turbines for highest steam conditions.

justified and if a higher capital cost can quickly be paid for in fuel savings. A typical example is the world's first 1000 MW reheat steam turbine at Ravenswood power station built in the early 1960s. This cross-compound unit features HP and IP turbines driving a 3600 rpm generator and three LP turbine sections with 40-in. last-stage blades driving a 1800

30 ● 100 YEARS OF POWER PLANT DEVELOPMENT

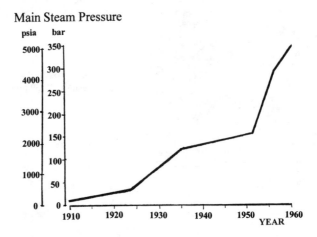

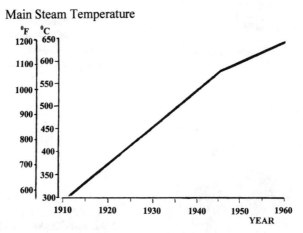

FIGURE 2-14. Main stream pressure and temperature trend towards super critical conditions.

rpm generator, as illustrated in Figure 2-15.[11] The unit was designed for the moderate steam conditions of 166 bar/538°C/538°C (2400 psia/ 1000°F/1000°F). All five turbine sections are designed as double-flow turbines. Cross-under piping connects the IP turbine exhaust with the inlets of the three LP turbine sections. Two boiler feedwater pumps are fluid-coupled to each end of the 3600 rpm shaft arrangement and a DC-exciter is gear-driven by the 1800 rpm generator.

An example of a thermal cycle of a typical 1960 vintage reheat steam turbine power plant is illustrated in the form of a heat balance diagram

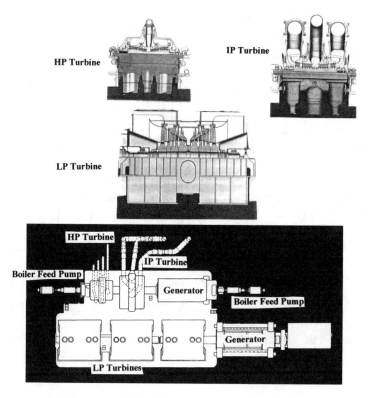

FIGURE 2-15. 1000 MW steam turbine at Ravenswood Power Station.

in Figure 2-16. The heat balance diagram provides the information about the thermal cycle layout and documents the thermodynamic performance of the steam turbine. A seven-feedwater heater cycle is shown with three LP turbine steam extractions and two IP turbine extractions. The two remaining feedwater heaters are supplied with steam from the HP and IP turbine exhaust. The drains from the HP feedwater heaters cascade down to the de-aerator and the drains from the LP feedwater heaters cascade down into the condenser. Depending on the feedwater system design, other more efficient drain systems can be considered. With the final feedwater heater connected to the cold reheat line, the final feedwater temperature is defined by the HP turbine exhaust (cold reheat) steam pressure and the delta temperature of the final feedwater heater. For supercritical pressure cycles a higher feedwater temperature

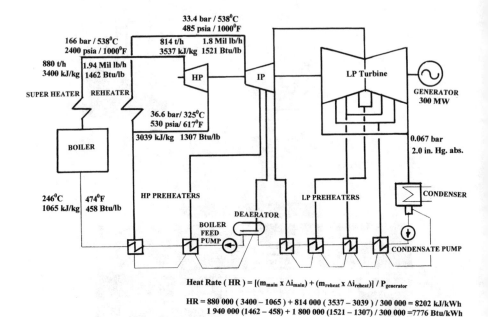

FIGURE 2-16. Heat balance diagram of reheat steam turbine.

is desired and can be provided by an additional steam extraction of the HP turbine.

With the final feedwater, the main and reheat steam conditions, and the generator output, the heat rate and efficiency of a steam cycle can be calculated as follows:

$$HR = m_{main} \times \Delta i_{main} + m_{reheat} \times \Delta i_{reheat} / P_{generator}$$

$$\Delta i_{main} = i_{main\ steam} - i_{final\ feedwater}$$
(enthalpy rise from final feedwater to main steam)

$$\Delta i_{reheat} = i_{hot\ reheat} - i_{cold\ reheat}$$
(enthalpy rise from cold to hot reheat)

The heat rate calculation in the heat balance diagram shows a heat rate (HR) of 8202 kJ/kWh or 7776 Btu/kWh. This result can also be expressed as an efficiency value:

$$\eta_{\text{heat balance}} = 3600/8202 \text{ kJ/kWh} = 3412.14/7776 \text{ Btu/kWh} = 43.9\%$$

In case the heat rate is given in kcal/kWh, the efficiency calculation would be:

$$\eta_{\text{heat balance}} = 860/1959 \text{ kcal/kWh} = 43.9\%$$

This efficiency is the steam cycle efficiency including the turbine and generator efficiencies; it does not include boiler and plant auxiliary losses. Adding these losses provides the power plant overall efficiency:

$$\eta_{\text{power plant}} = \eta_{\text{heat balance}} \times \eta_{\text{boiler}} \times \eta_{\text{auxiliaries}}$$

$$\eta_{\text{power plant LHV}} = 43.9\% \times 88\% \times 95\% = 36.7\%$$

The boiler efficiency, which includes pipe system losses, is based on the low heat value (LHV) of coal. The difference between the low and high heat value (HHV) is different for each kind of coal. A typical value is 5%, which lowers the power plant efficiency based on HHV to:

$$\eta_{\text{power plant HHV}} = 36.7\% \times 95\% = 34.9\%$$

This steam turbine power plant efficiency was achievable in the 1950s with the conventional steam conditions shown in the heat balance diagram. Since then improvements of all power plant components have been made to raise this plant efficiency to a 39% level based on HHV. Figure 2-17 shows that with cycle improvements this level could be raised to 42.8% based on HHV, equal to 45% based on LHV, and even higher.[12] Proven technology is available to build such highly efficient coal-fired steam turbine power plants. However, economical considerations (coal price versus capital cost) might not justify building such plants with all the measures shown. Some of the steps or part of the steps illustrated could be applied.

The improvement of the steam turbine efficiency level from 60% to 90% took about 90 years and involved a large number of steps. This evolutionary process was not only a result of design improvements of the turbine flow path, but also an outcome of the improved steam conditions. Early steam turbines operated mainly below the saturation line

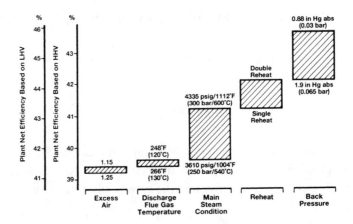

FIGURE 2-17. Measures to improve pulverized coal-fired power plant efficiency.

and with a high moisture content in the last stages. Modern reheat units with a high reheat steam temperature operate mainly in the superheated region and have only a low moisture content at the last-stage rows of their LP sections.

Two-dimensional flow path design, laboratory testing, and performance testing in power plants have already led to very efficient reheat steam turbines in the 1950s. New impulse-type as well as reaction-type blade profiles were developed and tested in multiple stage test turbines before being built into prototype steam turbines.

The later development in steam turbine design to reach efficiency levels above 90% is illustrated in Figure 2-18, revealing turbine component improvements made to reach this goal.[12] Improvements were possible in all turbine sections, with the major advances being related to the improvement of the turbine blade path. These latest steam turbine efficiency improvements are the result of the application of three-dimensional design tools. Three-dimensionally designed blade profiles are utilized in all steam turbine sections from the HP to the LP turbine. Today, integrally shrouded blades, as illustrated in Figure 2-19, are machined with numerically controlled machine tools, forming the three-dimensional profile.[13]

With the three-dimensional design, the reaction characteristics of the blades change from stage to stage and over the blade length of each stage

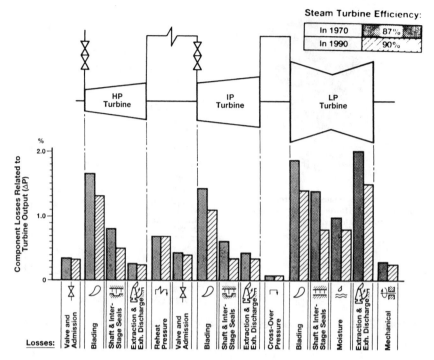

FIGURE 2-18. Steam turbine component improvements to achieve a 90% efficiency level.

to achieve the best possible performance of the entire turbine flow path. Proper design of the blade profile includes the transition from the blade root and the transition to the blade shroud. With this three-dimensional design the debate of impulse-type blading versus reaction-type blading loses its basis.

In addition, the three-dimensional design of steam turbine inlet and exhaust sections provides an overall turbine efficiency improvement. Also, improved shaft and inter-stage seal designs have been developed to reduce steam leakage losses. LP turbines with longer last-stage blades made from chromium steel alloys or titanium provide a larger annulus area per LP turbine flow. As shown in Figure 2-20, large tandem-compound steam turbines rated at 1000 MW are built with these newly developed design features and operate as supercritical units with advanced steam conditions.[13]

36 ● 100 YEARS OF POWER PLANT DEVELOPMENT

Three dimensional integrally shrouded blades produced by CAD-design to numerically controlled manufacturing

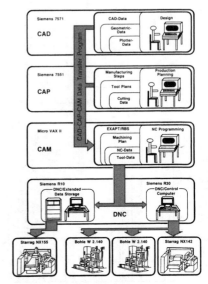

CAD-CAP-CAM-DNC Configuration
for Integrally Shrouded Blade Production

FIGURE 2-19. Three-dimensional designed and machined blades.

Throughout the design and construction process of these large, advanced, and highly efficient and reliable steam turbines, a variety of failure modes had to be dealt with. The major areas of failure modes are noted below. Design criteria and concepts have been established to address these issues and to avoid such failure modes in the future.

- **Rotor Instability.** Some of the larger size steam turbines with relatively slender rotors experienced instability at full-load operation. Steam whirl was identified as the cause. By building turbines with relatively stiff rotors having short bearing spans and large rotor diameters, this failure mode has been avoided. In addition, bearing designs have improved to provide more dampening and higher stability limits.[14,15]
- **High-Cycle Fatigue.** High-cycle fatigue turbine blade cracking has occurred. However, proper blade design eliminated this problem. Partial-arc admission first-stage blading was exposed to high dynamic stresses. Adopting full-arc admission avoids such high dynamic stressing. Proper dampening of potentially high vibra-

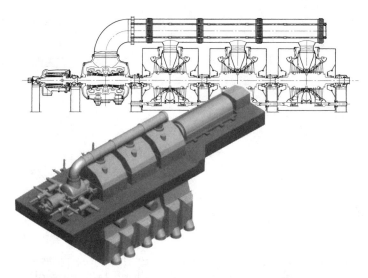

FIGURE 2-20. Tandem compound 1000 MW size steam turbine.

tions of all blade rows eliminates blade vibration causing high-cycle fatigue.[16] The large LP turbine blades must be properly tuned to avoid excess vibrations caused by blade or blade system resonance with the rated or a multiple of the rated speed. Such proper tuning is illustrated in Campbell diagrams of the last three stages of an LP turbine with free-standing blades in Figure 2-21.[17] Experience has shown that proper tuning avoids resonance, especially of low-frequency blade vibration modes with low-frequency speed harmonics.[18]

- **Low-Cycle Fatigue.** Load cycling and start-up procedures, as well as controls, have been developed to avoid excessive component thermal stressing and, consequentially, low-cycle fatigue. Operating a full-arc admission HP turbine in a variable pressure mode does not only achieve the highest efficiency performance, but equally important, eliminates thermal stressing due to load cycling as illustrated in Figure 2-22.[19] A load swing from full load to 33% part load with a partial-arc admission HP turbine results in a temperature change of 94°C (169°F) in the first stage turbine section when operating in a nozzle-controlled mode. Utilizing a full-arc admission HP turbine reduces this temperature change

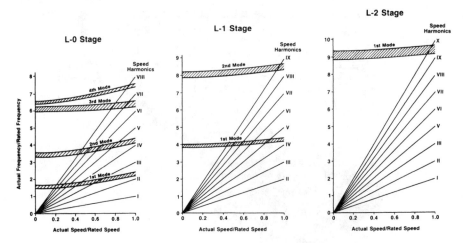

FIGURE 2-21. Campbell diagrams of the three rows from an LP turbine with 750 mm (30 in.) long last-stage blades.

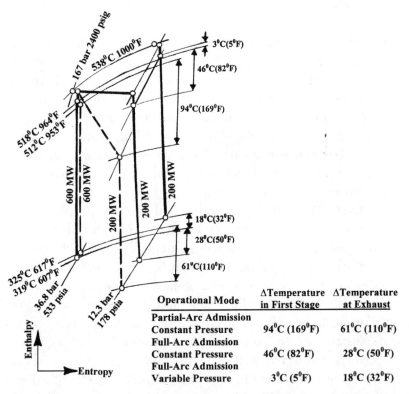

Operational Mode	ΔTemperature in First Stage	ΔTemperature at Exhaust
Partial-Arc Admission Constant Pressure	94°C (169°F)	61°C (110°F)
Full-Arc Admission Constant Pressure	46°C (82°F)	28°C (50°F)
Full-Arc Admission Variable Pressure	3°C (5°F)	18°C (32°F)

FIGURE 2-22. HP turbine temperature changes due to load swings from 600 MW full load to 200 MW part load.

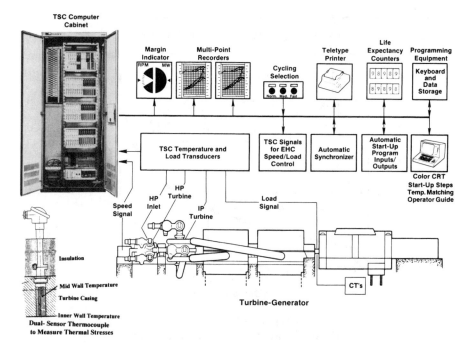

FIGURE 2-23. Turbine stress and start-up control system of the 1970s.

in the first stage to 46°C (82°F) when operating like the partial-arc admission turbine with constant pressure. Changing the operation mode to variable pressure operation further reduces the temperature change in the HP turbine first stage to only 3°C (5°F), which shows that thermal stressing due to load swings has been eliminated. As shown in this figure the temperature changes in the HP turbine exhaust section indicate similar trends toward lower changes in temperature levels.

For IP turbine sections of reheat steam turbines low-cycle fatigue is also a concern. Since IP turbines typically feature larger rotor diameters than HP turbine rotors, their thermal stresses during start-up have to be controlled.

In the 1950s, the first digital controls were applied for start-up systems to control thermal stresses in steam turbine power plants. Such systems today are updated to computerized start-up systems. However, Figure 2-23, of a 1970s vintage system, shows

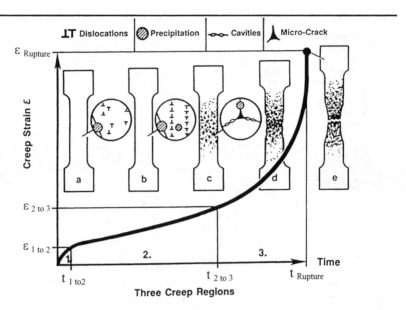

FIGURE 2-24. Creep strain curve with three regions and resulting microstructure changes.

the various functions that are expected from such a system, from selecting the operation mode to component life expectancy recording.[20] It is important to use actual metal temperatures and temperature differences in heavy component walls as real time data for thermal stress control.

Such systems can also be retrofitted into existing units. Retrofitting is important in cases where the mode of operation will be changed from base load to two-shift or peaking operation.

- **Creep.** With increasing metal temperatures of steam turbine components above 400°C (750°F), the creep rupture strength of materials has to be taken into consideration as design criteria. High-temperature materials were developed to keep creep effects within acceptable limits. As shown in Figure 2-24, creep is generally acceptable in its early two phases, but puts lifetime limits on high-temperature components depending on the changes of the material's microstructure. After the second creep phase has past, the formation of cavities and microcracks requires a replacement of high-temperature turbine components.[21]

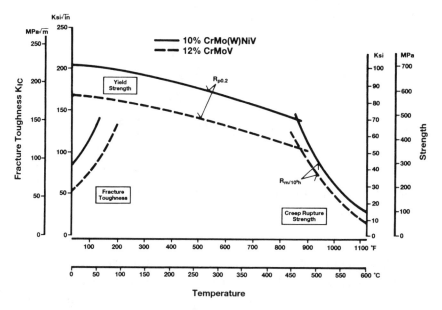

FIGURE 2-25. Creep strength, yield strength, and fracture toughness of advanced and conventional rotor forgings.

The latest metallurgical developments have led to the introduction of forgings and castings referred to as super-clean materials. Figure 2-25 compares an advanced 10% chromium steel forging with a conventional 12% chromium steel forging.[22] The advanced forging shows much improved properties. The creep rupture strength at about 600°C (1100°F) metal temperature is roughly twice as high as the creep rupture strength of the conventional forging material. At the same time the yield strength and fracture toughness have been drastically improved. Advanced materials are used today for components of new, large steam turbines, but could also be used in case of a creep problem with existing units.

- **Stress Corrosion.** Stress corrosion has always been a concern in power plants. It is influenced by three factors, as illustrated in Figure 2-26. Stress corrosion must be expected with inadequate water chemistry, insufficient material properties, and high tensile stress levels. Measures have been taken to get stress corrosion under control by improving on all three influential factors.

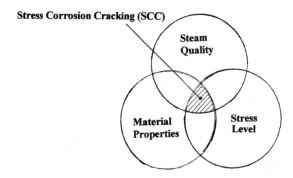

FIGURE 2-26. Three influential factors of stress corrosion cracking.

The water chemistry of a power plant's water/steam cycle is of utmost importance. The major source of impurities comes into the cycle by cooling water leakage. Condenser tubes and condenser tube-to-tube shield connections have been the main cause of leakage. The condenser design, assembly, and materials selection have improved to reduce the potential of cooling water leaking into the condensate. With the introduction of once-through boilers that do not have a drum to collect and flush out impurities, there have been initial problems with carryover of impurities in the main steam resulting in deposits and stress corrosion. Today, it is recommended for once-through boiler units to utilize larger condensate polishing systems (preferred 100% size) to ensure a high purity of the steam. It is recommended to keep the conductivity of condensated steam measured downstream of a strong acidic cation exchanger below 0.2 μS/cm (μmho/cm) during normal operation, and below 0.5 μS/cm (μmho/cm) during startup. Another impurity source is air intake. LP turbine components have shown stress corrosion due to a high CO_2 content (more than 300 ppb) carried into the steam/water cycle by air. Depending on the water treatment, the oxygen level in the steam/water cycle can be quite different, however air intake is a concern with all treatments because of the possibility of impurities such as carbon dioxide (CO_2) being carried into the system.[22,23]

Stress corrosion cracking is influenced by the kind of material that has been applied. For example, it has been realized that

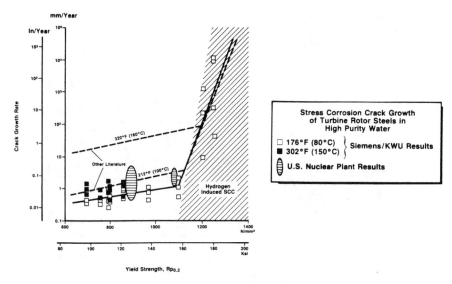

FIGURE 2-27. LP turbine rotor and disk material stress corrosion crack growth versus yield strength.

LP turbine components with high yield strength levels are susceptible to stress corrosion cracking. Figure 2-27 shows results from various sources (tests and power plant experience) that illustrate how LP turbine rotor and disk materials are more susceptible to stress corrosion if they are heat treated to achieve high yield strength levels.[24] The diagram shows a threshold at about 1100 N/mm^2 (160 ksi) yield strength at which the stress corrosion crack growth rate increases rapidly due to hydrogen-induced stress corrosion cracking. Below the threshold the influence of the materials yield strength level is relatively small.

High tensile stressing is the third factor needed for stress corrosion cracking. This stressing can be just a local stress raiser at a component surface. Based on extensive testing and long-term operating experience, the relationship of tensile stressing and environmental conditions has been well established, as illustrated in Figure 2-28.[25] At a tensile stress level of 50% of the material's yield strength level, no stress corrosion is expected even under severe corrosion conditions of the environment. If the environment is condensing steam, stress corrosion is only expected when

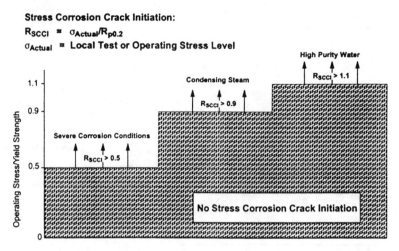

FIGURE 2-28. Stress corrosion crack initiation of LP turbine rotor and disk materials.

the tensile stressing is 90% of the material yield strength. Testing in high purity water showed that stress corrosion initiation did not occur at tensile stressing below 110% of the yield strength. However, the designer must realize that power plants do not operate with absolutely pure water and the operator should at all times keep the water chemistry within the defined limits.

Steam turbine components, especially LP turbine blading, that operate at or close to the saturation line or Wilson line are susceptible to stress corrosion cracking since impurities of the steam separate in this region and deposit on component surfaces. To avoid stress corrosion cracking one must always keep in mind that stress corrosion is already occurring if the three influential factors are locally present.

As we will see in the next chapter, stress corrosion became a major issue for nuclear power plants, and, extensive knowledge has been gained from fossil and nuclear power plant operation. The present knowledge of the stress corrosion phenomena allows us to properly design power plant components for eliminating stress corrosion cracking.

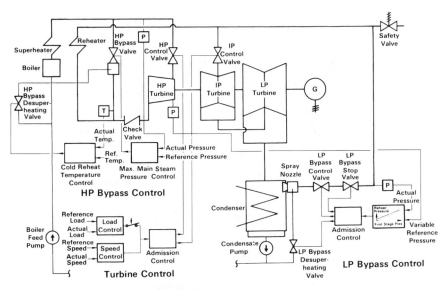

FIGURE 2-29. Reheat steam turbine plant with external main and reheat bypass systems.

- **Erosion.** Solid particle erosion of mainly HP turbine first-stage blades and droplet erosion of mainly LP turbine last-stage blades has occurred. However, measures have been established to mitigate these problems. Solid particle erosion can be related to particles that have not been flushed out of the steam/water cycle before startup by steam generator cleaning and proper steam blowing. Corrosion products can also do harm. A way of minimizing solid particle erosion is to utilize HP turbine full-arc admission and to operate in a variable pressure mode. External bypass systems, as illustrated in Figure 2-29, have proven to bypass solid particles during startup, and therefore minimize solid particle erosion. The HP bypass feeds steam into the cold reheat line and the LP bypass dumps hot reheat steam into the condenser. With this arrangement steam flows are established in the main and reheat steam systems, bypassing any particles around the HP and IP turbine sections before starting the turbine.[26]

Droplet erosion of last-stage moving blades is caused by slow moving water droplets being hit by the fast-rotating blade tip as

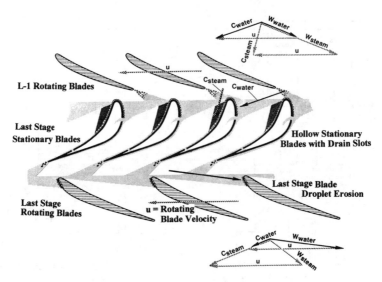

FIGURE 2-30. Droplet erosion reduction with hollow stationary blades.

illustrated in Figure 2-30. To minimize the amount of droplets leaving the last-stage stationary blades' trailing edges, these hollow blades have been equipped with drain slots at the suction and pressure side of the blade profile.[27] An internal heating system of the hollow stationary last-stage blades, as shown in Figure 2-31, is another way of mitigating droplet erosion by avoiding droplet formation on the last-stage stationary blade surface.[13] For protection of last-stage rotating blades against droplet erosion, their leading edges are protected by stelite strips or flame hardening of the leading edges.[17]

The development of highly efficient and reliable steam turbine plants was an evolutionary process over a period of 100 years. Today, steam turbines for even the most advanced steam conditions can be built for reliable long-term power plant operation with a minimum of maintenance effort.

The early coal-fired power plants were not very efficient or environmentally friendly. Smoking factory and power plant stacks were a symbol of prosperity. With the improvement of power plant performance, specific emissions were also drastically reduced. The power plant effi-

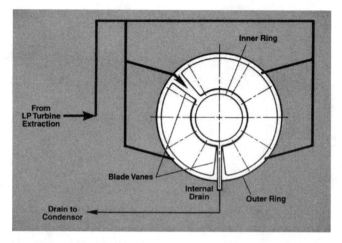

FIGURE 2-31. Steam heating of stationary last-stage blades.

ciency improvement from 15% to 45% alone accounts for a 200% reduction in most of the specific pollutant discharge per kilowatt-hour of generated electric power. The first major improvement to reduce pollutant discharge was the introduction of dust removal from the boiler flue gas. Today, electrostatic precipitators can be more than 99.9% effective in separating the combustion dust from the flue gas before it is discharged into the atmosphere.

The next step was the introduction of scrubbers to mitigate sulfur dioxide (SO_2) emissions. Not only were new power plants equipped with scrubbers, but older power plants were retrofitted with such systems. Figure 2-32 shows an example, and the drastic emissions reduction, of a lignite-fired power plant in which wet scrubbers were retrofitted and the original stacks eliminated. The cleaned flue gas is now discharged into the atmosphere through the cooling towers with much improved distribution patterns.[28]

For the reduction of nitrogen oxide (NO_X) emissions the combustion systems, especially the burners designed for pulverized coal combustion, were improved and, as an additional measure for bituminous coal-fired plants, catalytic converters were installed. Catalytic converters for NO_X emission reduction are generally designed for about 90% effectiveness as an economically justifiable layout. As shown in Figure 2-33, bituminous coal-fired power plants with catalytic converters reach a NO_X

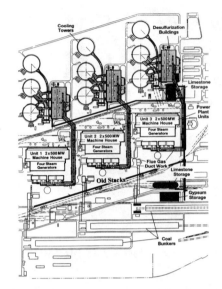
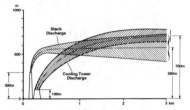

Discharge of flue gas through cooling towers

	Present	Goal 1996/97	Reduction
Dust			
Metric Tons	12,500	4,700	60%
Short Tons	13,800	5,200	
SO_2			
Metric Tons	301,000	36,200	88%
Short Tons	331,000	39,800	
NO_x			
Metric Tons	35,800	17,900	50%
Short Tons	39,400	19,700	

Expected emission reduction

FIGURE 2-32. Emission reduction including retrofit of wet scrubbers at a 6 × 500 MW lignite-fueled power station.

emission level that is about 70% above the NO_X emission of combined-cycle power plants burning natural gas without catalytic converters installed. Emissions are typically given in parts per million (ppm) of the flue gas. However, for a real comparison of the emissions of different power plants it is important to use the specific emission based on the actually generated kilowatt-hour of electric power.

The next major issue for the power generation industry could be the discharge of carbon dioxide (CO_2). The specific carbon dioxide discharge

- **Coal-Fired Power Plant without Catalytic Converter**
  6 g/kWh (13 x 10^{-3} lb/kWh) NO_X Discharge

- **Coal-Fired Power Plant with Catalytic Converter of 90 % NO_X Removal Rate**
  600 mg/kWh (1.3 x 10^{-3} lb/kWh) NO_X Discharge

- **Combined Cycle Power Plant without Catalytic Converter**
  350 mg/kWh (0.8 x 10^{-3} lb/kWh) NO_X Discharge

FIGURE 2-33. NO_x emission of coal-fired power plant with and without catalytic converter.

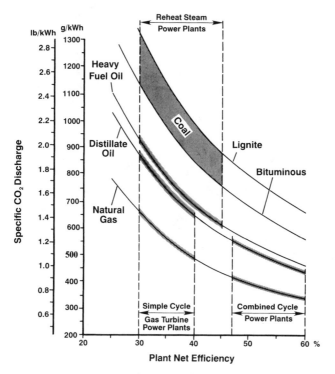

FIGURE 2-34. Specific carbon dioxide discharge as function of power plant efficiency and kind of fuel.

per generated kilowatt-hour of electricity is basically a function of the power plant net efficiency and amount of carbon dioxide produced by the combustion of different fuels. Figure 2-34 shows the specific carbon dioxide discharge of fossil power plants combusting different fuels at different efficiency levels.[1] It reveals the advantage of building a fossil-fired power plant with a high plant efficiency; it also shows that burning natural gas instead of coal for electric power generation provides a major reduction in carbon dioxide. However, not shown in the figure is the fact that nuclear power plants do not discharge any of the discussed pollutants.

CHAPTER 3

Nuclear Steam Turbine Power Plants

The first nuclear reactors built in the 1940s and early 1950s were research facilities of the government defense program. In the 1950s the first electric power was generated in some of these nuclear plants. After their successful operation, commercial nuclear power plants were built utilizing mostly light-water reactors (boiling water and pressurized water reactors). The unit size of these power plants grew rapidly over the next 20 years, reaching a unit rating of 1300 MW in the early 1970s. This fast development included the design of large, half-speed, single-shaft turbines for operation with saturated main steam. Steam turbine designers had the advantage of 60 years of experience building similar steam turbines with increasing unit ratings for fossil power plants. However, the specific main steam flow in lb/kW or kg/kW of nuclear steam turbines is about twice as large as the main steam flow of fossil steam turbines. Because of their high ratings and large specific flows, all nuclear steam turbines, with the exception of few early units and some 50 Hz units, are built as half-speed turbines operating at 1800 rpm in 60 Hz grid systems and 1500 rpm in 50 Hz systems.

A typical example of a large, half-speed 1300 MW steam turbine is illustrated in Figure 3-1, featuring one double-flow HP turbine and two double-flow LP turbines.[29] The low operating speed allows for the design of these turbines with heavy blading and a large annulus area per LP turbine flow section that utilizes last-stage blade lengths of up to 50 in. for 60 Hz and 60 in. for 50 Hz turbines. Most nuclear steam turbines feature monoblock HP turbine rotors and LP turbine rotors with shrunk-on disks. Few LP turbine rotors feature welded disks, and only at a later stage did large monoblock half-speed LP turbine rotors become available.

The design criteria for nuclear turbines versus fossil turbines are somewhat different. Where fossil HP and IP turbine sections have to handle high steam temperatures and are therefore exposed to low-cycle fatigue and creep, the nuclear HP turbines have to handle a high moisture content in the steam and are therefore exposed to erosion-corrosion.

Figure 3-2 shows the expansion lines of a fossil reheat turbine and expansion lines of a non-reheat and a reheat nuclear steam turbine. The main steam pressure of nuclear turbines is, in most cases, saturated steam with a pressure of 70 bar (1000 psia) or less. The saturated main steam expands to the external moisture separator pressure level. At this

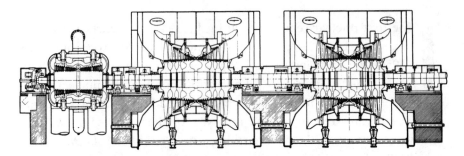

FIGURE 3-1. 1300 MW tandem compound nuclear steam turbine.

point the moisture content in the HP turbine exhaust steam is about 15%. In the case of a non-reheat unit, the 15% water is separated from the steam, and saturated steam or steam with a small moisture content enters the LP turbine sections. After expansion in their first stages, LP turbine internal drain systems become effective in separating some water out of the steam passing through the LP turbine's later stages. The expansion line endpoint without internal moisture separation shows a moisture content of more than 16%, the internal moisture removal reduces the moisture at the LP turbine exhaust to about 14%. However, with 14% moisture in the exhaust steam droplet erosion at the last-stage moving blades can be expected. With nuclear reheat steam turbines, the steam from the HP turbine exhaust first enters the external moisture separators and then the steam-heated reheaters. The moisture separator/reheaters (MSRs) reheat the HP turbine exhaust steam to about 260°C (500°F). With this steam temperature the LP turbine inlet conditions become similar to the LP turbine inlet conditions of a fossil reheat steam turbine. The moisture content of the exhaust steam is reduced to almost 10%, only about 1% higher than that of a fossil reheat steam turbine with internal moisture removal.

Nuclear reheat steam turbines have a higher efficiency than nuclear non-reheat steam turbines because their first LP turbine stages operate with super-heated steam and their later LP turbine stages are exposed to a much lower moisture content. In addition, the thermal cycle is improved, providing a 2% to 3% power plant efficiency increase. Another advantage of reheat is that it increases the LP turbine enthalpy drop and utilizes steam heating, which both reduce the LP turbine exhaust flow.

Nuclear Steam Turbine Power Plants ● 55

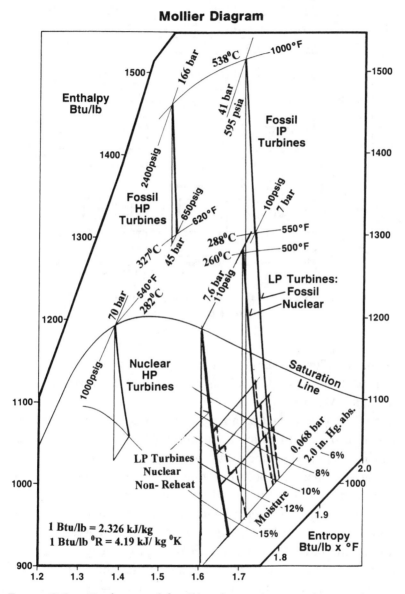

FIGURE 3-2. Nuclear and fossil/nuclear reheat and non-reheat steam turbine expansion lines.

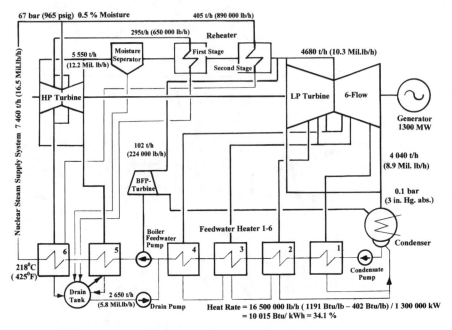

FIGURE 3-3. Heat balance diagram of a 1300 MW nuclear power plant.

Since the annulus discharge area determines the LP turbine size, the smaller exhaust steam flow allows for a smaller LP turbine to be built for the same MW rating.

Figure 3-3 shows a heat balance diagram of a 1970s vintage 1300-MW nuclear reheat steam turbine application. The thermal cycle features a two-stage reheat and six feedwater heaters. The moisture separator and two-stages of reheat are built into one unit. Two of these moisture separator/reheater (MSR) units are installed horizontally on each side of the LP turbine sections on the machine house floor. Proper drainage of the moisture separator and the reheater sections is achieved by utilizing a common drain tank. The performance of this turbine is evaluated by calculating the heat rate using the main steam plus the steam for the second stage reheat as the thermal cycle heat input. The 1300 MW are generated at a heat rate of 10 015 Btu/kWh or 34.1% plant efficiency. Today, more efficient units can be built to achieve a net power plant efficiency of about 37%. Utilizing MSRs together with the six feedwater heaters reduces the LP turbine exhaust to only 54% of the main steam.

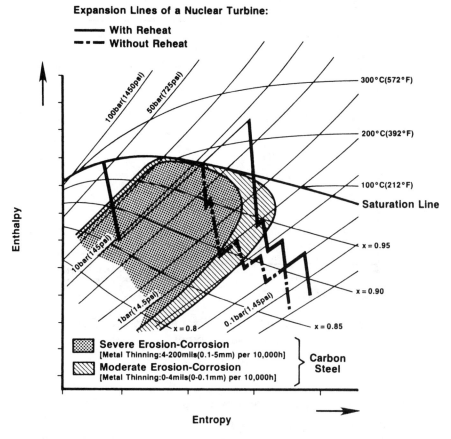

FIGURE 3-4. Operating conditions causing erosion-corrosion.

Proper moisture separator/reheater unit designs have been developed and built based upon experience. MSRs are exposed to extreme conditions causing erosion-corrosion, thermal stressing, and stress corrosion cracking. MSRs are available with single- and two-stage reheaters and can be arranged in a horizontal or vertical position. Selecting the proper material for each MSR component and designing adequate drainage of the moisture separator as well as the reheater systems mitigates erosion-corrosion effects and instability problems due to two-phase flow conditions.

Erosion-corrosion can be expected under the operating conditions indicated in Figure 3-4.[30] With increasing moisture content, the later flow section of the HP turbine is exposed to severe erosion-corrosion attack.

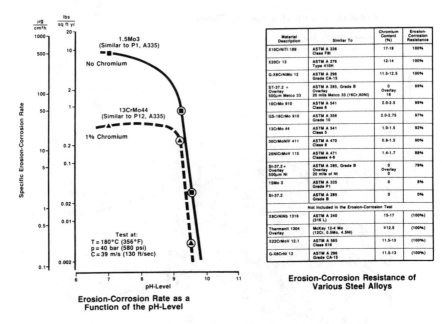

FIGURE 3-5. Criteria that affect erosion-corrosion in steam turbines.

Severe erosion-corrosion conditions are present in nuclear steam turbines with moisture separators only, from the HP turbine exhaust through the pipe system to the moisture separator and again in the first stages of the LP turbine. For nuclear steam turbines with moisture separator/reheaters (MSRs), the reheat and the first LP turbine stages are not exposed to erosion-corrosion, but some of the later LP turbine stages operate under conditions where moderate erosion-corrosion attacks can be expected. Severe erosion-corrosion did not occur in LP turbines of nuclear reheat units but did occur in LP turbines of non-reheat fossil and nuclear units.

Proper water treatment and proper material selection can minimize erosion-corrosion attacks. Figure 3-5 shows the effect of the pH level of the steam/water cycle on the specific erosion-corrosion rate and the material selection on the erosion-corrosion resistance. A pH level above 9.0 drastically reduces the erosion-corrosion rate. The resistance to erosion-corrosion is practically 100% with 12% to 17% chromium steel alloys. Changing carbon steel components with 2.5% chromium steel components increases the resistance from 0% to about 97% to 99%. The 0% of

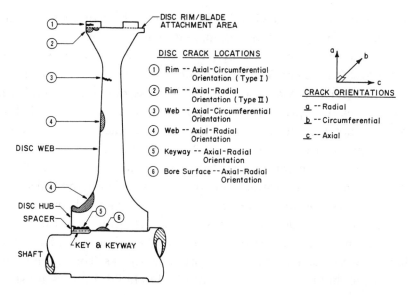

FIGURE 3-6. Stress corrosion cracks in LP turbine rotors.

carbon steel is the lowest resistance of a steel component; however, this high susceptibility of carbon steel components can be improved with a nickel coating to about 79% erosion-corrosion resistance.[30] The proper flow path design is also an important measure to mitigate erosion-corrosion by avoiding at any location high flow velocities of steam with high moisture content.

Stress corrosion became a more difficult issue after the Electric Power Research Institute (EPRI) reported in 1982 that stress corrosion cracking in nuclear power plant LP turbine rotors was found. This finding was related to power plant safety by the Nuclear Regulation Commission (NRC) due to the potential of turbine rotor parts (missiles) leaving the machine house and hitting the reactor building. Missile analyses were performed to establish LP rotor inspection intervals based on potential stress corrosion crack initiation and crack growth.

Figure 3-6 best summarizes the findings of EPRI.[31] Stress corrosion occurred on LP turbine rotors at various locations such as:

- blades and blade attachments
- rotor or disk rims
- rotor or disk fillets
- disk to shaft shrink fit
- key ways at shrink fit

The stress corrosion cracking found revealed shortcomings in water treatment, material selection, design, and manufacturing.

Steam purity was improved to keep the conductivity of condensed reheat steam below 0.2 μS/cm. The need to improve the water treatments of nuclear power plants was also required because of corrosion problems in nuclear steam supply systems (NSSS), such as stress corrosion in boiling water reactor piping and tube thinning, and cracking in steam generators of pressurized water reactors. With the measures recommended for the NSSS, the operating conditions for LP turbine rotors were at a level at which stress corrosion potential was sufficiently minimized.

Rotor and disk forgings were heat treated to have a yield strength level well below the earlier described threshold of 1100 N/mm^2 (160 ksi). Manufacturing and assembly of rotors often resulted in locally high tensile surface stresses causing local stress corrosion attacks. This has not only been avoided, but measures were introduced to build compressive stress into component surfaces.

A large variety of design modifications were adopted for LP turbine rotors of all suppliers to mitigate stress corrosion. The general goal was to reduce tensile stresses since stress corrosion does not occur without tensile stressing or at low tensile stress levels. Measures taken to reduce tensile stress levels of shrunk-on disks are shown in Figure 3-7, by building compressive stresses into the disk's surface, by a specifically developed heat treatment, by shot peening, and by keyway rolling.[32] As an example of improving a disk-type rotor design, a ten-disk rotor design advanced to an eight-disk, and finally to a six-disk design as illustrated in Figure 3-8.[33] With the relatively wide disk forgings the heat treatment for building compressive stresses into the disks' surfaces became more effective. These compressive stresses have been measured at a depth of 30 mm (1.25 in.) from the disk surface.

A large number of LP turbine rotors have been replaced with advanced designs mitigating stress corrosion. Rotor designs such as shrunk-on disk-type rotors, monoblock rotors, and welded rotors are available, which warrant an inspection period of at least 100,000 operating hours. LP turbine rotor replacement has been typically applied to nuclear turbines of the 1970s or earlier. In these cases, the enter LP turbine flow path was redesigned utilizing advanced blading that can improve the nuclear plant heat rate or efficiency by 4%. This also results in a 4% higher net output of a nuclear power plant without any change of the reactor

Nuclear Steam Turbine Power Plants • 61

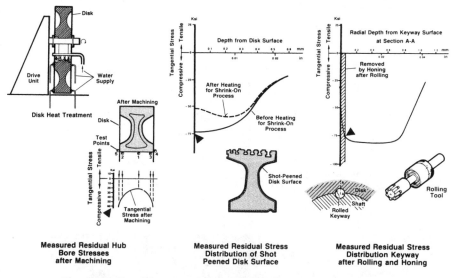

FIGURE 3-7. Measures to establish compressive surface stresses in rotor disks.

output. For a 1300 MW nuclear power plant, the increase in net output of 4% accounts for 52 MW, which in itself can already justify an LP turbine replacement.

Nuclear power plants have proven to be highly reliable in supplying base load power between refueling outages. The refueling outages over the years have become shorter with record outage times of only three weeks. Refueling takes place in one to one-and-a-half year intervals and depends mainly on the unit's capacity factor.

The clear advantage of nuclear power plants is that an almost unlimited fuel supply is available and specific fuel costs per generated kilowatt-hour are low. Most important is the fact that nuclear power plants do not discharge any dust, sulfur dioxide, nitrogen oxide, carbon monoxide, or carbon dioxide.

Unfortunately nuclear power plants are expensive, and the licensing process and construction take a long time. Also there are the safety and long-term fuel storage concerns that have become political issues in many countries. The political effect on the nuclear power issue is best revealed by comparing the nuclear power industry in France and Germany. Both countries have been very successful in building nuclear power plants.

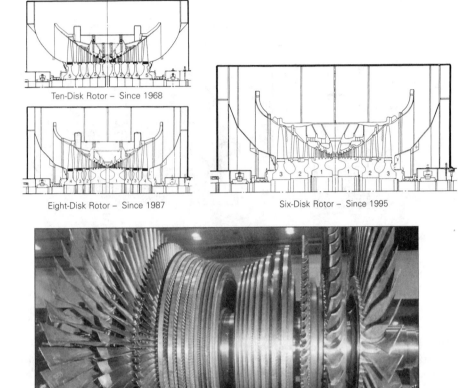

FIGURE 3-8. Disk-type LP turbine rotor development from a ten-disk to a six-disk rotor design.

However, the nuclear power generation industry activities in Germany came to a complete stop because of political opposition, whereas in France the nuclear industry is in full swing and 75% of the electrical power in France is generated by nuclear power plants. In Germany, it has been decided to close down all nuclear power plants. It has been reported that Russia is considering building 30 nuclear power plants at their western border to supply electricity to the deregulated Western

European market. Maybe a global or international approach, especially for the nuclear power generation industry, would be more meaningful.

Nuclear power plants have been in operation for more than 30 years. The Chernobyl accident in Russia should not be used against nuclear power plants in general. Such an accident is practically impossible with reactors built in the United States or Western Europe. In addition, reactor designs for power generation have been greatly improved, and even inherently safe reactor designs are available today.

Presently, the political climate against nuclear power and the high costs of nuclear power plants has slowed the growth of the nuclear industry. However, as soon as the concern about emissions and greenhouse effects caused by power generation intensifies, safe nuclear power plants will become the answer. Another issue that could reverse the present slowdown of the nuclear industry could be a shortage of low-cost natural gas for power generation.

When nuclear power generation moves to the forefront, further improved power plant concepts should be adopted. Fast breeder and high-temperature gas-cooled reactor designs are already available. If there is an economic need for more efficient nuclear power plants, high-temperature gas-cooled reactors could be built that will feature a combined-cycle design with a gas turbine cycle followed by a heat recovery steam generator/steam turbine as bottom cycle.

Co-Generation Plants

Co-generation is almost as old as power generation. Even in the early years of power generation with steam turbines there was the desire to utilize steam turbine extraction or backpressure steam for industrial processes as well as for commercial and domestic heat supplies. Gas turbines, because of their large amount of high-temperature gas discharge, were always candidates for co-generation. The heat of the high temperature exhaust gases can be recovered in a heat recovery steam generator to build a combined-cycle power plant, a co-generation plant, or the combination of both. The major co-generation plant concepts with gas and steam turbines, which have been applied, are illustrated in Figure 4-1.

Gas turbines with their high temperature turbine exhaust gas are able to supply high-level heat for any kind of process. Instead of discharging the high temperature gas through the stack, a heat recovery system can be installed upstream of the stack. This kind of co-generation is especially attractive with midsize and small gas turbines in industrial applications. This co-generation concept has no effect on the gas turbine except that it increases the gas turbine backpressure. Therefore, one should keep in mind that a 1% output of a heavy-duty, full-size gas turbine is lost when raising the backpressure by roughly 20 mbar or 8 in. H_2O. There are also limits given for a maximum allowable backpressure of gas turbines.

The trend of the power generation industry toward deregulation and selective distributed power generation provides an opportunity for small industrial electricity, steam, and/or heat users to build small co-generation plants. Lately, the term *combined heat and power* (CHP) is used for these small co-generation plant concepts. Today, a large variety of aeroderivative and heavy-duty gas turbines in the 1 kW to 70 MW output range are available to provide the optimal match of the specific electricity and steam needs. Packages like the one shown in Figure 4-2 are offered as just gas turbine-generator packages with all their auxiliaries or as turn-key packages that include the heat recovery steam generators with their auxiliary systems.[34] The example features an aeroderivative gas turbine co-generation plant, but small heavy-duty gas turbines are also available. Although the simple-cycle efficiency of smaller gas turbines is lower, their fuel utilization to generate electricity and deliver steam can make a small facility very attractive.

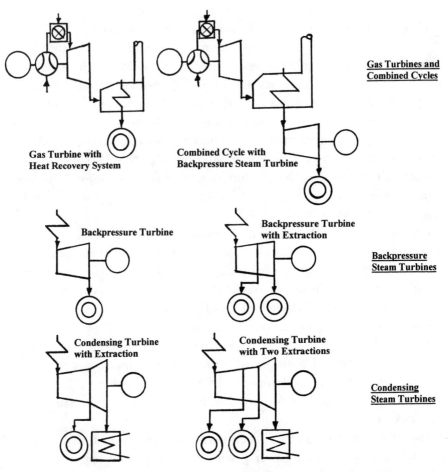

Figure 4-1. Co-generation plant concepts with steam and gas turbines.

To increase the electric power generation efficiency of such small plants, they can also be built as combined-cycle co-generation plants. Co-generation with combined-cycle power plants provides a wide variety of plant concepts with different gas and steam turbine designs. Figure 4-1 shows one example of a combined-cycle co-generation plant utilizing the gas turbine discharge heat in a heat recovery steam generator (HRSG) to supply steam for a backpressure steam turbine.

Backpressure steam turbines utilize the entire turbine steam flow as a heat source after expanding in the turbine. In cases where steam with different pressure and temperature conditions is needed, backpressure

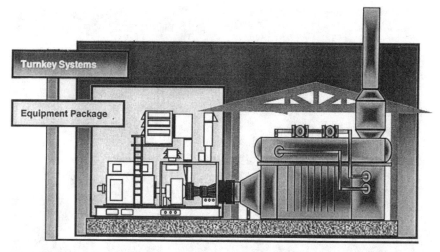

Figure 4-2. Aeroderivative gas turbine package for co-generation.

turbines can also be built with one or more extractions. The advantage of backpressure turbines is that they provide the highest fuel utilization. Their disadvantage is that electric power generation is limited and a direct function of the steam needed.

Condensing steam turbines with extractions eliminates this disadvantage by providing an LP turbine section to produce additional electric power. This LP turbine section can also be designed for optional operating without supplying any extraction steam for co-generation. The number of co-generation extractions can be one, two, or even more.

For steam turbine and combined-cycle power plants not only the shown examples of steam turbine designs can be applied, but even more design concepts are available. Very early in the power generation history it was realized that co-generation becomes more attractive if the steam turbine extraction pressures can be controlled. The first steam turbine with such automatic control features, shown in Figure 4-3, was already built in 1907.[6] Today, turbines with multiple extraction controls are available and all the co-generation large steam turbines can also be built as reheat steam turbines.

Co-generation plants provide the best fuel utilization since the highest losses of any steam turbine or combined-cycle power generation plant, namely the heat rejection in the condenser, are either eliminated or minimized. However, finding the right way of evaluating their performance

FIGURE 4-3. First steam turbine with pressure-controlled extraction built in 1907.

is difficult. The following examples show that electrical power generation and heat production must be recognized as two different types of energy supplies. The selected examples are one steam cycle with a condensing turbine and two steam cycles with backpressure turbines. The main steam conditions of the three cycles that will be compared are the same, namely:

Main steam pressure	69 bar	1000 psia
Main steam temperature	538°C	1000°F
Main steam flow	454 mt/h	1 000 000 lb/h

Figure 4-4 shows the three non-reheat steam turbine expansion lines, the first being a condensing turbine and the second and third being expansion lines of two backpressure turbines with different efficiencies:

Expansion Line	Turbine Efficiency	Turbine Backpressure
Condensing turbine (1)	82%	0.068 bar (2 in. Hg. abs.)
Backpressure turbine (2)	83.2%	5.2 bar (75 psia)
Backpressure turbine (3)	70%	5.2 bar (75 psia)

In the evaluations of all three cases the following efficiencies have been assumed:

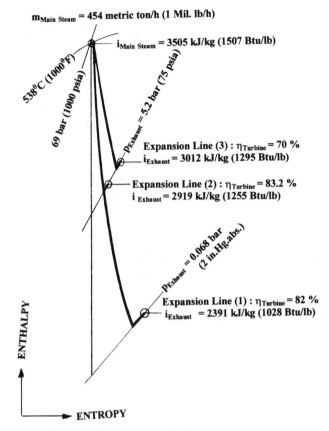

FIGURE 4-4. Expansion lines of condensing and backpressure turbines.

$\eta_{\text{Generator}} = 97\%$ (including mechanical losses)

$\eta_{\text{Auxiliaries}} = 98\%$

$\eta_{\text{Boiler}} = 86\%$ (including boiler feed pump)

The calculations of power plant performance are based on the simplified steam/water cycles illustrated in Figure 4-5 without feedwater heaters. Based on the different backpressures of the condensing turbine and the two backpressure turbines, the feedwater has a different feedwater temperature and enthalpy. The feedwater enthalpy of the condensing turbine is 160 kJ/kg (69 Btu/lb) and the enthalpy of the two backpressure turbines 647 kJ/kg (278 Btu/lb).

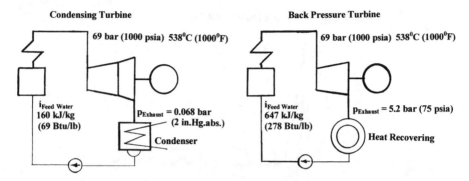

FIGURE 4-5. Heat balance diagrams of condensing and backpressure turbines.

The condensing turbine generates the maximum electric output with the assumed main steam at the following efficiency level:

$$P(1)_{Turbine} = m_{Main\ Steam} \times (i_{Main\ Steam} - i_{Exhaust})/3412$$

$$= 1\ 000\ 000 \times (1507 - 1028)/3412 = 140\ 400\ kW$$

$$P(1)_{Power\ Plant} = P(1)_{Turbine} \times \eta_{Generator} - \eta_{Aux}$$

$$= 140.4 \times 0.97 \times 0.98 = 133.5\ MW$$

$$Q(1)_{Input} = m_{Main\ Steam} - (i_{Main\ Steam} - i_{Feedwater})/\eta_{Boiler}$$

$$= 1\ 000\ 000 \times (1507 - 69)/0.86 = 1672\ Mil\ Btu/h$$

$$HR(1) = Q(1)_{Input}/P(1)_{Electric}$$

$$= 1\ 672\ 000\ 000/133\ 500 = 12\ 520\ Btu/kWh$$

$$\eta(1) = 3412/HR(1) = 3412/12\ 520 = 27.3\%$$

This result shows that the given main steam has been used for power generation only, generating in the condensing turbine an output of 133.5 MW with a plant efficiency of 27.3%.

The same main steam will now be used for co-generation with a backpressure turbine. The electrical power will be less but the same main steam heat energy will provide a much better fuel utilization. The electrical power is generated at the following efficiency.

$$P(2)_{Turbine} = 1\,000\,000\,(1507 - 1255)/3412 = 73\,900 \text{ kW}$$

$$P(2)_{Power\ Plant} = 73.9 \times 0.97 \times 0.98 = 70.2 \text{ MW}$$

$$Q(2)_{Input} = 1\,000\,000 \times (1507 - 278)/0.86 = 1429 \text{ mil Btu/lb}$$

$$HR(2) = 1\,429\,000\,000/70\,200 = 20\,360 \text{ Btu/kWh}$$

$$\eta(2) = 3412/20\,360 = 16.8\%$$

With the same main steam, only 70.2 MW (instead of 133.5 MW) are generated at an efficiency of only 16.8% (instead of 27.3%); however, a very large amount of heat energy is utilized accounting for a very high fuel utilization. The total co-generation heat is the entire steam flow multiplied by the backpressure turbine exhaust enthalpy minus the enthalpy of the returned feedwater, which can be expressed in thermal MWs:

$$Q_{Co\text{-}Gen}(2) = m_{Exhaust} \times (i_{Exhaust} - i_{Feedwater})$$

$$= 1\,000\,000 \times (1255 - 278) = 977 \text{ mil Btu/h}$$

$$= 977\,000\,000/3412 = 286\,300 \text{ kW}$$

Together with the electrical power the total utilized heat energy can be expressed as:

$$Q_{utilized}(2) = 286\,300 + 70\,200 = 356\,500 \text{ kW}$$

Now a heat rate and efficiency of the co-generation plant can be calculated based on this utilized heat energy of 356.5 MW, instead of the electric output of only 70.2 MW, to be:

$$FU(2) = Q(2)_{Input}/Q_{Utilized}$$

$$= 1\,429\,000\,000 / 356\,500 = 4008 \text{ Btu/kWh}$$

or

$$\eta_{FU}(2) = 3412/4008 = 85.1\%$$

It is important to distinguish this result as heat energy utilization of the fuel for a co-generation plant and to use the term *fuel utilization* (FU) instead of the term *heat rate* (HR). The advantage of co-generation can best be seen by comparing this result with the first example of a condensing turbine cycle in which heat rate and fuel utilization are the same, and therefore the efficiency $\eta_{FU}(1)$ is also 27.3% versus the 85.1% of the backpressure co-generation unit. However, a comparison of the efficiency of the two cycles in regard to generating electric power shows the 27.3% versus 16.8% in favor of the condensing turbine cycle.

Proper evaluation of co-generation is a difficult task since electric power generation and heat supply take place at different energy levels; however, they directly affect each other. The third example has been selected to reveal some of these difficulties. It uses the same cycle conditions as the second example, but the backpressure turbine has an efficiency of only 70% per expansion line (3). The output of this less efficient turbine is only:

$$P(3)_{Turbine} = 1\,000\,000 \times (1507 - 1295)/3412 = 62\,100 \text{ kW}$$

$$P(3)_{Power\ Plant} = 62.1 \times 0.97 \times 0.98 = 59.0 \text{ MW}$$

The heat rate and efficiency for generating electric power are:

$$Q(3)_{Input} = Q(2)_{Input} = 1429 \text{ mil Btu/h}$$

$$HR(3) = 1\,429\,000\,000 / 59\,000 = 24\,220 \text{ Btu/kWh}$$

$$\eta(3) = 3412/24\,220 = 14.1\%$$

The co-generated heat is:

$$Q_{Co\text{-}Gen}(3) = 1\,000\,000 \times (1295 - 278) = 1\,017 \text{ mil Btu/h}$$

$$= 1\,017\,000\,000/3412 = 298\,100 \text{ kW}$$

The utilized heat amounts to

$$Q_{Utilized}(3) = 59.0 + 298.1 = 357.1 \text{ MW}$$

and the fuel utilization is:

$$FU(3) = 1\,429\,000\,000/357\,100 = 4002 \text{ Btu/kWh}$$

$$\eta_{FU}(3) = 3412/4002 = 85.3\,\%$$

The results of Cases (2) and (3) reveal the importance of determining different performance evaluation factors when comparing two co-generation cycles. The 70% turbine efficiency of Case (3) versus 83.2% of Case (2) certainly results in a much lower electric output and efficiency. However, the fuel utilization shows a move in the other direction with a better performance of the lower turbine efficiency Case (3).

	Electric Output and Efficiency	Fuel Utilization
Case (2)	70.2 MW and 16.8%	85.1%
Case (3)	59.0 MW and 14.1%	85.3%

The reason for the improved fuel utilization of Case (3) is due to the fact that the loss of electrical output is caused by a lower turbine efficiency, which results in a higher turbine exhaust enthalpy and is, therefore, recovered as co-generation heat. Since this heat is assumed to be supplied without losses (while the turbine output is not only affected by the turbine efficiency but also by the generator and auxiliary losses), the fuel utilization becomes larger and the efficiency better. Using the fuel utilization as an overall co-generation efficiency indicates that the best co-generation cycle is a cycle without electric power generation:

$$Q_{Max} = m_{Main\ Steam} - (i_{Main\ Steam} - i_{Feedwater})/3412$$

$$= 1\,000\,000 \times (1507 - 278)/3412 = 360\,200 \text{ kW}$$

$$FU_{Max} = Q_{Input}/Q_{Max} = 1\,429\,000\,000/360\,200 = 3967 \text{ Btu/kWh}$$

$$\eta_{FU} = 3412/3967 = 86\%$$

which of course represents the boiler efficiency.

Another very misleading approach to evaluating co-generation heat is an assumption that the provided heat is sold and therefore evaluated as supplied heat:

$$Q_{Co\text{-}Gen} = m_{Co\text{-}Gen} \times i_{Co\text{-}Gen}$$

This approach applied to Case (2) would show the following:

$$Q_{Co\text{-}Gen} = 1\,000\,000 \times 1255/3412 = 367.8 \text{ MW}$$

$$FU = HR(2) \times P(2) / [P(2) + Q_{Co\text{-}Gen}]$$

$$FU = 20\,360 \times 70.2 / (70.2 + 367.8) = 3263 \text{ Btu/h}$$

$$\eta_{FU} = 3\,412/3263 = 105\%$$

which certainly is an overly optimistic result.

The most efficient co-generation cycle is best identified by the cycle that generates the most electric output from the given cycle heat input and with the given co-generation heat demand.

There is an endless number of co-generation cycles that have been developed. Steam turbines are specifically designed to match any application. Turbines used for co-generation are also referred to as industrial turbines. Backpressure turbines are often designed as high-speed units, leading to very small and efficient turbine designs. Condensing turbines can become very large, especially co-generation turbines that have been built for large district heating plants.

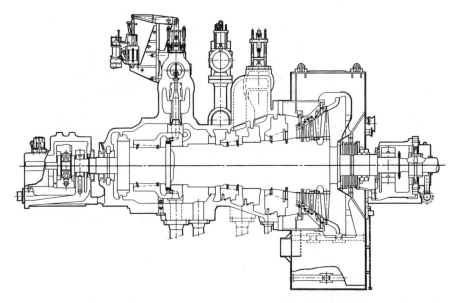

FIGURE 4-6. Industrial-type steam turbine with dual extraction.

The following two examples, both extraction steam turbines with two controlled extractions, reveal the wide range of steam turbine designs that have been developed. Figure 4-6 depicts a typical industrial steam turbine design for about 60 MW output.[35] The single-casing turbine features four turbine building blocks and three sets of control valves. The first set of valves controls the main steam flow to the turbine and the remaining two sets of valves control the extraction pressure for supplying co-generation steam at two different pressure levels. The first HP blade group section or HP building block is arranged as reverse-flow to balance its thrust forces with the thrust forces of the remaining three blade groups. The second IP blade group expands the steam from the first to the second extraction pressure level. The two remaining blade groups are the LP turbine section for condensing operation to the condenser vacuum. Such turbine design provides a wide operating range with steam being provided at controlled pressure levels that are adjustable in a certain range; condensing operation without steam extractions is also possible. This concept of combining building blocks into one single casing turbine allows one to design extraction turbines as well as backpressure turbines for each specific application. The valve sets and building blocks

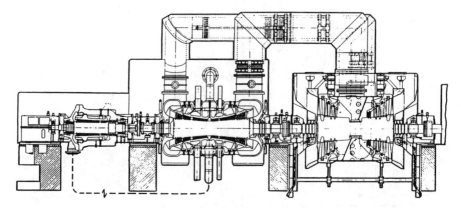

Figure 4-7. Reheat steam turbine with dual extraction for district heating.

are available in different sizes and for different pressure and temperature ranges to suit most industrial turbine applications.

The second steam turbine illustrated in Figure 4-7 is of a different design: a reheat steam turbine that has been designed as a multiple extraction turbine.[36] This kind of turbine has been designed for up to a 400 MW rating as reheat steam turbines for district heating. They feature a HP turbine, an IP turbine, and an LP turbine casing. The HP turbine is a single-flow design to expand the main steam to cold reheat steam conditions. The IP turbine is a double-flow turbine with two large extraction ports at each side for steam extractions at two different pressure levels: one steam extraction for providing hot water and one at a lower pressure and temperature level for providing district heat. In addition, steam extractions at other pressure levels provide process steam, steam for feedwater heating, and additional steam for the hot water supply. The blade paths of the two IP turbine flows are different with the design pressure levels of one IP turbine exhaust being 4 bar (58 psia) and 1.15 bar (17 psia) of the other exhaust. The two LP turbine flow sections are also designed for different steam conditions. The steam flows from the two IP exhaust sides are separately admitted into the two separate LP turbine flow sections. The two IP as well as the two LP turbine flow sections are of different designs to best perform under the different volumetric steam flow conditions. The two extraction pressure lev-

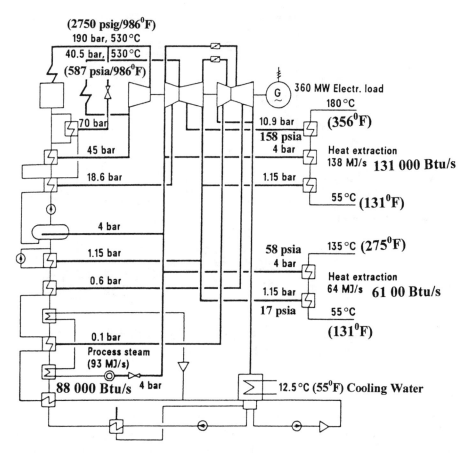

FIGURE 4-8. District heating plant with reheat steam turbine.

els are controlled by butterfly valves that are installed in each of the two separate cross-over pipes from one of the two different IP turbine exhausts to one of the two LP turbine inlets.

Figure 4-8 shows the heat balance diagram of a district heating plant that provides 93 MJ/s (88 000 Btu/s) process heat, 138 MJ/s (131 000 Btu/s) hot water, and 64 MJ/s (61 000 Btu/s) heat, in addition to 360 MW output. The unit can generate 410 MW when operating without steam extraction for co-generation. The heated water temperatures of 180°C (365°F) and 135°C (275°F) can be controlled by the butterfly valves in the two cross-over pipes, since their positioning controls the pressure levels in the two IP turbine sections.[36]

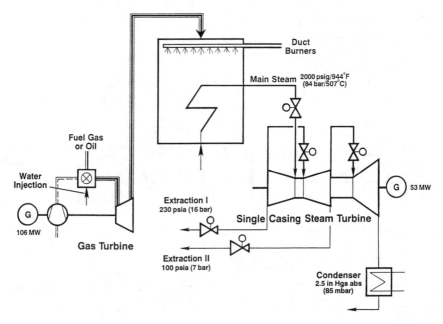

Figure 4-9. Combined-cycle co-generation plant with dual-extraction condensing turbine.

The following two examples are combined-cycle co-generation units. The first combined-cycle power plants designed as a co-generation unit is illustrated in Figure 4-9.[35] A dual-extraction steam turbine is applied in combination with a 106 MW gas turbine. The exhaust heat of the gas turbine is utilized in a heat recovery steam generator (HRSG) to generate main steam with 84 bar (1200 psia) pressure and 504°C (944°F) temperature for the 53 MW steam turbine.

The two extraction pressures of 16 bar (230 psia) and 7 bar (100 psia) can be controlled. If the steam supply from the HRSG is not sufficient, duct burners are installed in the HRSG to generate additional steam to satisfy any need for extraction steam and steam turbine output.

The second example, illustrated in Figure 4-10, is a district heating plant with three 63-MW gas turbines with three HRSG and one 73-MW backpressure steam turbine.[35] The steam turbine's main steam conditions are 80 bar (1150 psia) and 500°C (932°F). It features three extraction points for providing process steam with 16 bar (232 psia) pressure as well as district heating and hot water. In addition, main steam and

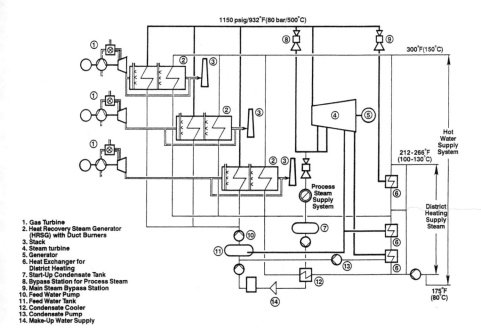

FIGURE 4-10. Combined-cycle district heating plant with three gas turbines.

hot water coming directly from the HRSGs can be utilized to supply process steam as well as district heating and hot water. The cycle layout provides the highest operating flexibility for this co-generation plant.

There are an unlimited number of examples with unique design features available to best provide co-generation. The excellent heat utilization of co-generation facilities makes them very attractive. Also, from an environmental point of view, the idea not to discharge heat into the atmosphere before it is fully utilized is a very desirable concept.

CHAPTER 5

Gas Turbine Power Plants

Large heavy-duty gas turbine power generation plants are low-specific capital cost plants with short construction times, from signing a contract to commercial operation in less than a year. Such plants consist of just a gas turbine-generator put on a foundation slab surrounded by its auxiliaries. A typical example is illustrated in Figure 5-1 and features an advanced 170 MW size gas turbine with a peaking plant (simple cycle) efficiency level of 38%.[37]

To develop large, heavy-duty advanced gas turbines with such high rating and efficiency was an evolutionary process over several decades. Early gas turbine peaking plants with heavy-duty gas turbines had an efficiency of only 20%. Gas turbine peaking plants with aeroderivative engines typically performed a couple of percentage points better than the same size heavy-duty units. Today, however, aero-engine technology has been adopted to build also more efficient heavy-duty gas turbines.

The first gas turbine design concept was invented and built about 100 years ago by Franz Stolze in Berlin, Germany. He applied for a patent for his fire turbine in March 1873. This was eleven years before Charles Parsons applied for a similar patent in the UK in 1884. The gas turbine of Franz Stolze's power plant, shown in Figure 5-2, was supposed to drive four generators with transmission belts.[38] The four generators were installed in the eaves of the machine house. The single-shaft, two-bearing gas turbine features an axial compressor and turbine section. The combustion system generated hot combustion air by burning coal. Franz Stolze also indicated that gaseous and liquid fuels could be burned in an internal combustion ring. There is proof that the gas turbine power plant built in 1904 as shown in Figure 5-3, but there is no proof that the plant ever generated any positive power output.[38]

The design concept of single-casing, single-shaft gas turbines with axial compressors has generally been adopted for today's large aeroderivative engines and heavy-duty gas turbines. There have been other developments in the 1900s to the 1950s with gas turbines featuring separate compressor and turbine housings, as well as gas turbines with radial compressors for industrial use. In the steel industry blast furnace gas was utilized as fuel for some gas turbines. However, the power generation industry in general only applied single-casing and single-shaft aero-engines and heavy-duty gas turbines as they became available at a later date. In the 1930s and 1940s such gas turbines were developed as air-

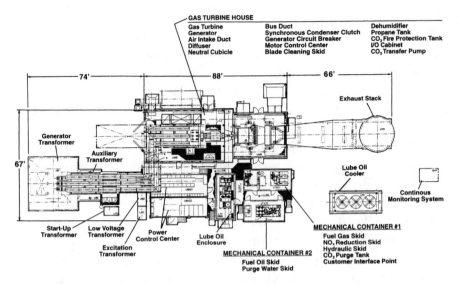

FIGURE 5-1. Full-size gas turbine peaking power plant layout.

craft jet engines. The first production models of aero-engines with axial compressors and internal combustion systems developed in Europe and the United States are illustrated in Figure 5-4.[39,40] Mass production of jet engines began in the mid-1940s, with about 6000 units of the Jumo design built in one year.

A large number of peaking power plants with aeroderivative gas turbines and some heavy-duty gas turbines were first used in the 1960s. At the same time the first combined-cycle power plants were introduced. The aeroderivative engines were more advanced and efficient than the first heavy-duty gas turbines. However, the demand for larger unit ratings led to the fast development of larger size and more efficient heavy-duty gas turbines. The best indication for this progress is the increase in the firing temperature of increasingly larger heavy-duty gas turbines as illustrated in Figure 5-5. The trend of the firing temperature shows a typical characteristic of an evolutionary process with an initially fast increase followed by a relatively small increase in the later years. However, a major step in developing advanced heavy-duty gas turbines was taken around 1990 when the first advanced gas turbines were introduced. The goal set by the Department of Energy (DOE) was to build highly efficient, combined-cycle power plants with a net power plant efficiency of

Gas Turbine Power Plants • 87

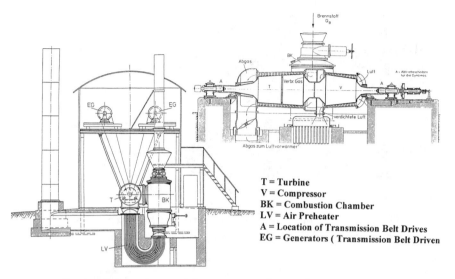

T = Turbine
V = Compressor
BK = Combustion Chamber
LV = Air Preheater
A = Location of Transmission Belt Drives
EG = Generators (Transmission Belt Driven

FIGURE 5-2. Gas turbine power plant of Franz Stolze.

FIGURE 5-3. Photograph of first gas turbine
with Franz Stolze in 1904.

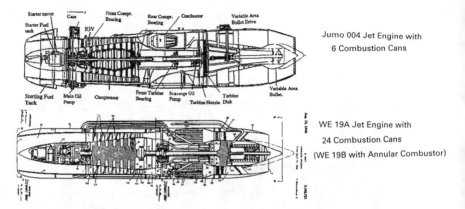

FIGURE 5-4. First jet engines with multiple-stage axial compressors designed and built in Germany and the US.

60%. Today, firing temperatures of heavy-duty gas turbines have reached about 1300°C (2400°F), and higher firing temperatures up to and even in excess of 1400°C (2550°F) can be expected.

The simple-cycle or peaking gas turbine plant cycle is a Brayton process, as shown in the temperature/entropy diagram of Figure 5-6. Air with an ambient temperature (A) is compressed to the combustion sys-

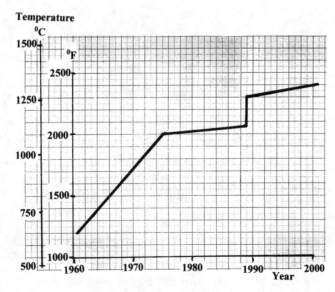

FIGURE 5-5. Firing temperature trend of heavy-duty, full-size gas turbines.

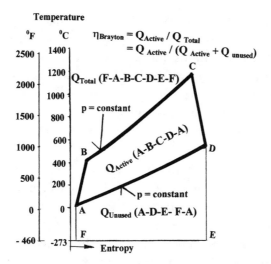

FIGURE 5-6. Simple cycle gas turbine Brayton process.

tem pressure level (A-B). Compression of the air raises the temperature and moves the air to a slightly higher entropy level, which represents the compressor efficiency. In the combustion system, the air temperature is further increased (B-C) to achieve the gas turbine inlet temperature level at point (C). From this level the combustion gas expands in the turbine section to the gas turbine exhaust conditions to be discharged into the atmosphere (C-D). The expansion causes a slight increase in entropy, which is determined by the turbine section efficiency. Because the turbine is powered with combustion gases, the general term gas turbine has been established in the power industry, regardless of the fuel being liquid or gas.

Hot exhaust gases are discharged into the atmosphere at (D) and ambient temperature air is taken from the atmosphere at (A); the process is then closed by the heat rejection into the environment from point (D) to point (A). Since this heat is rejected at a high temperature level, a large variety of co-generation and combined-cycle plant concepts have been developed and applied for recovering this heat and to minimize the heat discharge into the atmosphere.

The Brayton cycle efficiency can be calculated by dividing the two areas:

$$\eta_{Brayton} = Q_{Active}/Q_{Total} \triangleq A\text{-}B\text{-}C\text{-}D\text{-}A/F\text{-}A\text{-}B\text{-}C\text{-}D\text{-}E\text{-}F$$

Advanced heavy-duty gas turbines built as simple-cycle units burning natural gas achieve a power plant net efficiency of 38% or better, based on standard (ISO) conditions and low-heat value (LHV). The cycle efficiency can further be improved by increasing the firing temperature that moves up point (C), lowering the ambient temperature, which lowers point (A), or increasing the pressure ratio that moves up point (B). However, the most efficient combined-cycle design is achieved at a relatively low-pressure ratio.

All the efficiency information in this book, if not otherwise stated, is based on the low-heat value (LHV) of the used fuel. When comparing gas-fired power plants with plants burning other fuels, one should keep in mind that the high-heat value (HHV) of natural gas is typically 11% higher than its low-heat value (LHV). The difference between HHV and LHV is about 6% for #2 fuel oil and about 5% for bituminous coal. Heat rates are therefore higher and efficiencies lower by these percentages when basing the net plant efficiencies on high-heat values (HHV) of their fuels.

Since aeroderivative gas turbines typically have a higher pressure ratio than heavy-duty gas turbines, they provide a better cycle efficiency in simple-cycle applications. Today, advanced heavy-duty, full-size gas turbines are designed for a firing temperature level of about 1300°C (2400°F) and have adopted some aero-engine technology. Figure 5-7 illustrates present advanced heavy-duty, full-size gas turbines and their basic design concepts.[41,42,43] For example, the first advanced gas turbine shown was originally developed in the 1980s as a 150 MW advanced 7F unit. Based on operating experience and further development, it became a 7FA unit with increased output and improved efficiency. These, about 180 MW size, advanced gas turbines represent a technology referred to as *"F"* technology. Their pressure ratio is below 17:1, specifically designed for best performance in combined-cycle power plant application. Even more advanced and larger heavy-duty gas turbines referred to as *"G"* and *"H"* technology are in their prototype start-up phase. In developing such advanced gas turbines a number of issues had to be addressed:

- thermodynamic performance improvement
- materials for high-temperature turbine blading
- coatings to control high-temperature corrosion

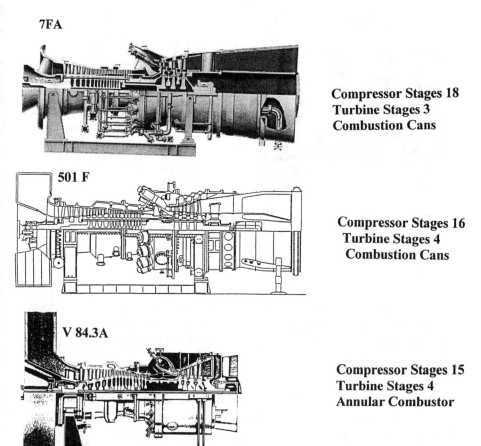

FIGURE 5-7. Three advanced full-size approximately 180 MW 60 Hz gas turbines.

- advanced turbine blade cooling to reduce metal temperature
- stable low-emission (dry low-NO_X) combustion at elevated firing temperature

The latest improvement in the thermodynamic performance of the compressor and turbine blade path designs was achieved by utilizing three-dimensional computer codes that were first used in designing the most efficient aero-engine blading. Early gas turbines featured cylindrical compressor blading, which has been replaced in advanced gas tur-

Cylindrical Blades

Three Dimensional
Blade Profile for Optimal
Flow Conditions over Its
Length Including Blade
Root and Tip Sections

FIGURE 5-8. Three-dimensional design of compressor blades.

bines by a three-dimensional blade design shown in Figure 5-8.[44] In addition, the advanced design tools allowed the design of blade root and tip sections for reduced boundary losses; also improvements in sealing designs were adopted.

Materials to handle high temperatures at high yield strength and high creep rupture strength were developed. Besides the development of nickel and cobalt alloys, which replaced high alloy steel as blade materials, the blade casting technology was improved. The precision casting technique was advanced to cast turbine blades that directionally solidified and even produced single-crystal blades, as illustrated in Figure 5-9.[45] These advances in casting techniques minimized the adverse effect of crystal boundaries with regard to reducing the material's high-temperature strength and high-temperature corrosion resistance. Single-crystal blading is available today for high temperature first- and second-stage turbine blade rows of even the largest heavy-duty gas turbines.

Not only does improved blade material allow for the operation at higher temperature levels, but also blade coating can provide the same effect. Two different kinds of coating have been developed, namely anticorrosion coatings and thermal barrier coatings. The application of such coatings to the first stages of turbine blading allows the firing temperature of gas turbines to be raised. The vacuum plasma spray (VPS) coating process with a variety of coating materials has proven to be an excellent anticorrosion coating for extremely high operating tempera-

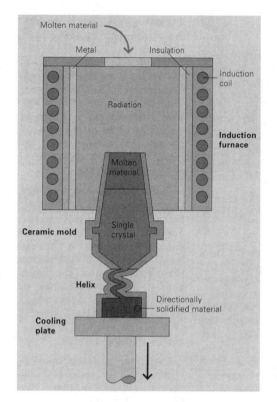 

FIGURE 5-9. Single-crystal first stage gas turbine blade.

tures. To keep high temperature gases away from the blade surface, thermal barrier coating is additionally applied. This ceramic coating is best applied with the electron beam physical vapor deposition (EB-PVD) process. The EB-PVD process forms rod-shaped zirconia (ZrO_2) crystallites as shown in Figure 5-10.[46] This thermal barrier coating has proven to provide long-term blade protection with relatively low coating wear.

Cooling techniques, such as film cooling and even more advanced cooling schemes utilizing a portion of the compressed air for blade cooling, were first developed for aero-engines. Figure 5-11 shows a first-stage stationary blade of a heavy-duty, full-size gas turbine with film cooling.[47] Cooling air is fed into the hollow blade and than passes through small holes to the blade's outer surface, forming a film of cool air. The cooling air film keeps the hot combustion air away from the blade surface and reduces the blade metal temperature.

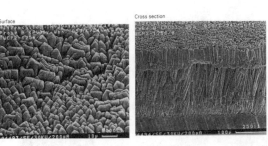
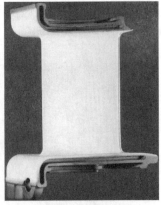

FIGURE 5-10. Electron beam physical vapor deposition (EB-PVD) coating.

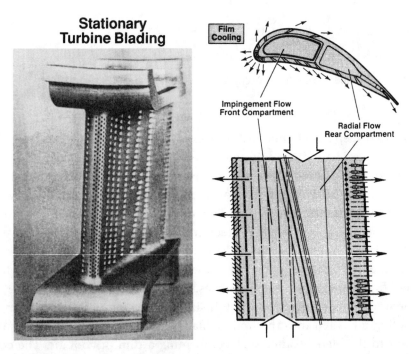

FIGURE 5-11. Film cooling of first-stage stationary blade.

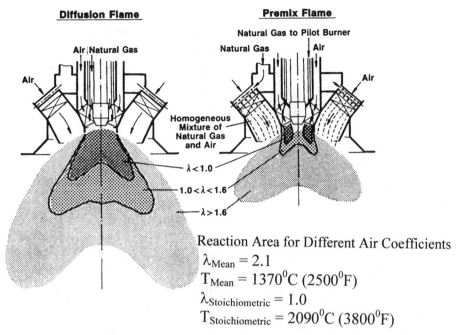

Reaction Area for Different Air Coefficients
$\lambda_{Mean} = 2.1$
$T_{Mean} = 1370°C\ (2500°F)$
$\lambda_{Stoichiometric} = 1.0$
$T_{Stoichiometric} = 2090°C\ (3800°F)$

FIGURE 5-12. Hybrid burner operation modes with natural gas.

Providing stable low emission required extensive research and development work including testing and real power plant operation. Dry low-NO_X emission combustion has been developed based on lean combustion in premix burners. Most efforts were related to natural gas combustion, but premix burners for fuel oil have been developed and are successfully in operation. When combusting a lean natural gas/air mixture the combustion temperature can be kept at about 1400°C (2500°F) with an air coefficient of $\lambda = 2.1$, which is much lower than the stoichiometric combustion temperature of 2100°C (3800°F) with the air coefficient of $\lambda = 1$. Figure 5-12 shows an example of a hybrid burner operating in the conventional diffusion burning mode and in the premix burning mode.[48] A small pilot flame is burning in the premix mode to ensure that the main premix flame cannot be extinguished due to transient operating conditions such as load rejections.

The development of dry low-NO_X emission combustion systems was not an easy task. Certain issues had to be addressed, such as premix flame flash back and premix flame combustion noise. Since fuel and air are premixed, they form an inflammable mixture before they enter the flame zone. Experience has shown that impurities in natural gas, such as oil or liquid hydrocarbons, as well as extreme transient conditions, can cause events of flashbacks. Combustion noise has been an issue with advanced gas turbines operating at higher and higher firing temperatures and with the demand for lower and lower dry low-NO_X emissions.[49] In most cases combustion noise occurs when operating at or near maximum load conditions of a gas turbine. A large number of measures have been taken and are still under development to extend the range of operating gas turbines with higher firing temperatures and to still achieve dry low-NO_X emissions. This includes the integration of catalytic combustion systems. Presently, heavy-duty gas turbines can be operated at dry low-NO_X emission levels in the range from single-digit to 25 ppm NO_X, depending on their firing temperature level.

As power plant test results in Figure 5-13 show, the dry low-NO_X emission can be reduced by simply reducing the mean flame temperature of a premix burner.[50] However, a reduction of the mean flame temperature below a threshold of about 1400°C (2550°F) leads to a steep increase of the CO emission. The optimal premix operation shown in the example was reached at an air coefficient range of λ = 2.0 to 2.2, when both the NO_X emission and CO emission reached values below 25 ppm.

Besides dry low-NO_X emission reduction, steam or water injection has also been utilized. Even though the premix of fuel oil has been successful in operation,[51] diffusion burners with water injection are used to typically keep NO_X emissions below 42 ppm. Water or steam injection into the combustion system is also used for power augmentation. When injecting the same amount of water and fuel oil into the combustion system, the gas turbine output increases by about 10%; however, the gas turbine efficiency is reduced by roughly 3.5%.

A way to raise the output and efficiency is to lower the compressor inlet temperature. Evaporation and fogging systems can reduce the compressor inlet temperature, as illustrated in Figure 5-14, from 35°C (95°F) to 25.6°C (78°F) by increasing the relative air humidity from 40% to 87%.[52] This results in a gas turbine output increase of about 7% and an efficiency increase of about 2%. In addition, wet compression can be utilized to further increase output and simple-cycle efficiency. Wet compression is limited by the amount of water, typically 2%, which can safely

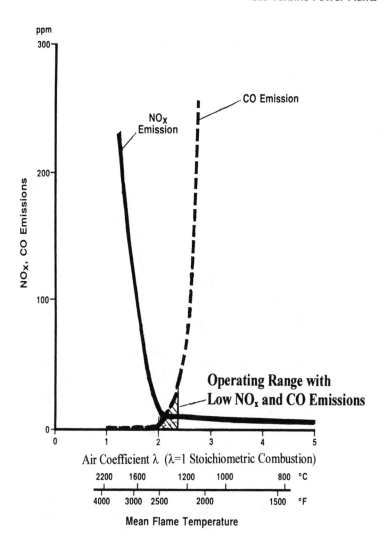

Figure 5-13. Dry low-NO$_X$ and CO emission data from power plant testing of premix burners.

be added to the compressor inlet air. Power augmentation can best be used at high ambient temperature levels because the gas turbine output is relatively low. With low ambient temperatures there are typically limits given in the form of a gas turbine's maximum output based on its mechanical design. For additional peaking capacity on high temperature

Example for Evaporative Cooler:
95°F Dry Bulb Temperature and 40% Relative Humidity gives a Wet Bulb Temperature of 75°F.

When using a Evaporative Cooler Effectiveness of 85% the Dry Bulb Temperature to be used as Compressor Inlet Temperature is Calculated as Follows:

95°F − (95°F − 75°F) x 0.85 = 78°F

The Relative Humidity can be found by using the Dry Bulb Temperature of 78°F where it crosses the Wet Bulb Temperature line of 75°F which gives 87% Relative Humidity.

Sea Level Diagram has been used which neglects the minor effect of altitude.

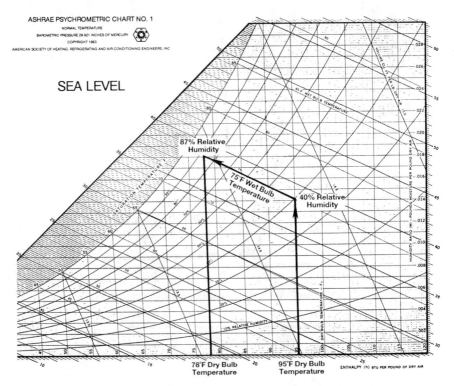

FIGURE 5-14. Compressor inlet conditions corrected for evaporative cooling.

summer days, chilling systems with ice storage can be applied to augment electric power.

In addition to the three gas turbines in Figure 5-7, an advanced heavy-duty, full-size gas turbine with dual or sequential combustion, as illustrated in Figure 5-15, is also available.[53,54] Like reheat of steam turbines,

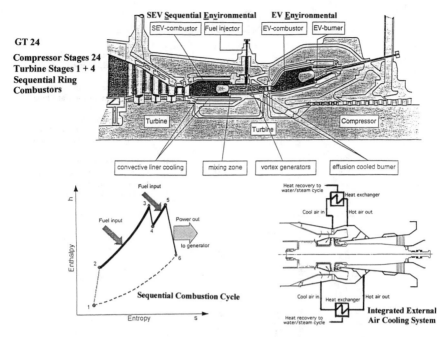

FIGURE 5-15. Advanced heavy-duty, full-size gas turbine with sequential combustion.

sequential combustion provides a thermal cycle improvement over single combustion. However, such a design concept also requires an increase in pressure ratio to about 30:1. The gas flow from the first combustion system (EV-combustor) is expanded in the first turbine stage from about 30 bar (435 psia) to about 15 bar (220 psia). After the second combustion in the SEV-combustor the gas expands in the remaining four turbine stages. Extensive cooling air is required at two different pressure levels to cool the high-temperature components of the two combustion systems. To apply this gas turbine design concept effectively, it is recommended for use in combined-cycle applications, where the relatively high heat losses of the two external air-cooling systems can be recovered.

The development of even larger and more efficient "G" and "H" technology gas turbines involves an increase in firing temperature to reach a 1420°C (2600°F) level and a rise in pressure ratio. The optimal pres-

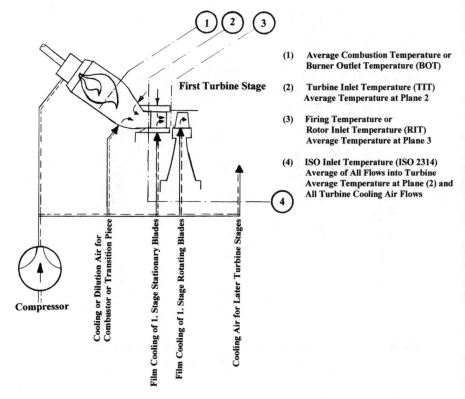

FIGURE 5-16. Definition of gas turbine temperatures.

sure ratio for simple-cycle and combined-cycle power plants is different and depends on the firing temperature and other factors.

Before discussing this development further, we should identify the different gas turbine inlet temperature definitions used in the industry. Figure 5-16 shows the four temperature definitions for the combustion and gas turbine inlet. The average combustion temperature or burner outlet temperature (1) reveals at what temperature combustion takes place. This temperature affects the dry low-NO_X emission level, which can be achieved.

The turbine inlet temperature measured upstream of the first-stage stationary blades in plane (2) can already be much lower than the combustion temperature depending on the use of dilution air or cooling air in an open loop to cool the transition section.

The firing temperature or rotor inlet temperature can be measured at plane (3) upstream of the first-stage rotating blading. This temperature is used in the United States to identify at which turbine inlet temperature level power is generated. It reveals to which temperature the first rotating blades and rotor components are exposed.

In Europe, typically the ISO turbine inlet temperature (4) is used. This temperature, per ISO 2314, is a calculated average temperature of all flows into the turbine, which is the average temperature at plane (2) and all of the turbine air flows. This calculated average temperature of, for example, an F class turbine with film cooling of the first-stage blading, is much lower than the firing temperature. Depending on the cooling air system applied, the ISO inlet temperature can be as much as 100°C (180°F) lower than the firing temperature. The ISO inlet temperature is the thermodynamic value that can be directly applied to calculate the gas turbine performance.

When extending the evolution of gas turbines into the future, it can be seen in the earlier Figure 5-5 that the firing temperature comes closer to the dry low-NO_X combustion temperature range of 1500°C to 1600°C (2700°F to 2880°F). This provides a limit for building more efficient gas turbines with high inlet temperatures and at the same time to provide dry low-NO_X emissions. Dry low-NO_X emission is achieved by premixing air with fuel and by combusting this lean mixture at a low temperature level. To provide high-efficiency advanced gas turbines with dry low-NO_X emission, design concepts have to be developed to bring these two temperatures, namely the combustion temperature and the firing temperature, closer together.

The relatively high temperature difference in the present advanced gas turbines is mainly a result of cooling requirements to reduce the metal temperature of the gas turbine inlet components. If high-temperature materials became available for these components to operate without cooling, the goal would have been achieved. However, the present materials and component designs need extensive cooling. Today, advanced gas turbines use air cooling for the first-stage stationary blades as open-loop film cooling and open-loop and/or closed-loop cooling for the combustion system.

One way to achieve high firing temperatures without increasing the combustion temperature is with the introduction of steam cooling. A variety of cooling methods utilizing steam as a cooling media have been

Transition Piece Cooling

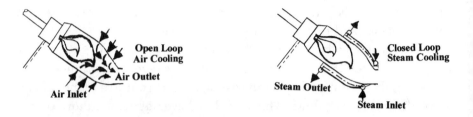

First Stage Stationary Blade Cooling

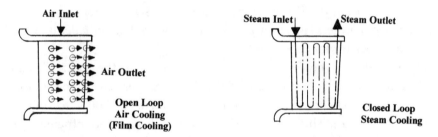

FIGURE 5-17. Steam cooling of transition piece and first stage stationary blade.

specifically developed for heavy-duty gas turbines. Since steam is available in combined-cycle power plants and the gas turbine cooling cycle can become an integral part of the heat recovery steam generator's steam cycle, steam cooling can improve the combined-cycle power plant performance. Steam cooling of gas turbine blading as well as hot gas components, like transition pieces, has been developed.

Figure 5-17 shows how steam cooling provides an increase in firing temperature of a gas turbine without raising the combustion temperature. The first example is a closed-loop steam cooling of a transition piece, which replaces the open-loop air cooling. With air cooling the average combustion temperature is lowered by roughly 110°C (200°F) before entering the first-stage stationary blades. With the introduction of transition piece steam cooling, this temperature drop can be minimized to about 40°C (72°F), which reduces the temperature difference between the combustion and firing temperature by 70°C (125°F). The second ex-

Transition Piece and First Stage Stationary Blade Cooling

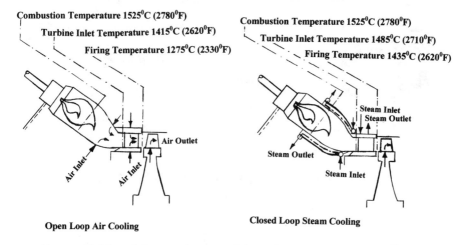

FIGURE 5-18. Advanced gas turbine air versus steam cooling.

ample reveals how the change from open-loop film air cooling of the first-stage stationary blades to a closed-loop steam cooling of these blades reduces the temperature difference across these stationary blades by roughly 90°C (160°F), from about 140°C to 50°C (250° to 90°F). In both examples open-loop cooling was replaced by closed-loop cooling. With open-loop air cooling, the relatively cold cooling air is mixed with the hot gas flow reducing its temperature. The closed-loop steam cooling system feeds the steam back to the steam cycle at an increased temperature level for the heat to be recovered.

The comparison of an air-cooled versus steam-cooled gas turbine is illustrated in Figure 5-18 for utilizing steam cooling of the transition pieces and first-stage stationary blades. The example is based on a combustion temperature of 1525°C (2780°F) for both cooling methods. The firing temperature with open-loop air cooling is 1275°C (2330°F), whereas closed-loop steam cooling could provide a firing temperature of 1435°C (2620°F). With this 160°C (290°F) higher firing temperature advanced gas turbines can be built to achieve a combined-cycle power plant efficiency of 60%, without raising the dry low-NO_X emission due to an increase in combustion temperature. However, the turbine components have to be designed to handle the higher firing temperature. Utilizing

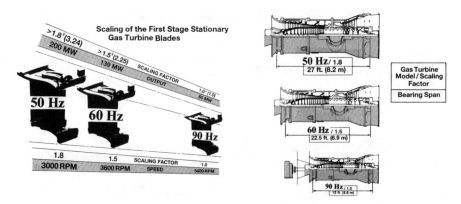

FIGURE 5-19. Scaling of gas turbines for different operating frequencies.

steam cooling for a simple-cycle application should only be considered if the heat energy of the steam can be utilized for co-generation and/or power augmentation.[55,56]

Besides the advanced full-size, heavy-duty gas turbines, there is a large variety of smaller gas turbines available for power generation. These turbines can be either heavy-duty or aeroderivative units. In the power generation industry there is no clear gas turbine size definition, but the terms mini- and micro-turbines are generally used. The following gas turbine size definition based on unit rating in simple-cycle applications could be used as a guide:

Gas Turbines	Large Full-size	Medium Mid-size	Small	Mini	Micro
60 Hz appl.	>100 MW	>10 MW	>1 MW	>50 kW	>1 kW
50 Hz appl.	>150 MW	>10 MW	>1 MW	>50 kW	>1 kW

The rating of 60 and 50 Hz units is the same except for the full-size units, because these units are typically scaled-up from 60 Hz to 50 Hz or scaled-down in size from 50 Hz to 60 Hz. An example of such scaling from 50 Hz to 60 Hz, and even to a mid-size 90 Hz gas turbine is shown in Figure 5-19. The scaling of the first-stage stationary blade represents the scaling technique for the entire flow path that keeps the stress levels and thermodynamic design, including temperature levels and pressure ratio of all three gas turbine models, at about the same level.

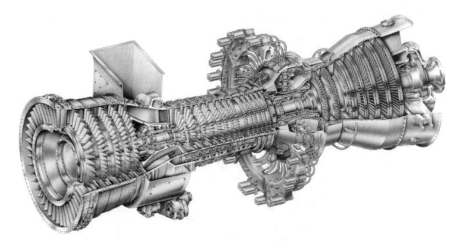

FIGURE 5-20. Aeroderivative mid-size gas turbine. (GELM600 with dry low-NO_x combustion system)

The linear scaling of the dimensions in accordance to the reciprocal frequency or speed of the gas turbines results in a change in their flows and outputs by the second power. The rating of the three units is therefore:

- 3000 rpm–200 MW
- 3600 rpm–139 MW
- 5400 rpm–60 MW

with the third unit being a mid-size heavy-duty gas turbine that can be utilized with gearboxes of different speed ratios to drive either a 3000 rpm or 3600 rpm generator. Scaling down and adding the gear losses reduces the performance of the mid-size unit by slightly more than the scaling factor.

An example of an aeroderivative mid-size gas turbine, shown in Figure 5-20, is a modified airplane engine with the turbine section driving a generator instead of providing thrust forces for an airplane. The highly efficient gas turbine is roughly a 42 MW unit, which achieves a simple-cycle power plant efficiency of around 41% with a pressure ratio of about 30:1. This LM6000 mid-size gas turbine is equipped with an annular dry low-NO_X combustion system to reach dry low-NO_X emissions below 25 ppm.[57] The dual-rotor design features a high-pressure 14 stage compressor driven by a two stage high-pressure turbine and a five-stage low-

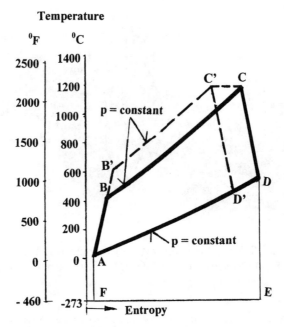

FIGURE 5-21. Effect of an increased pressure ratio on the gas turbine exhaust temperature.

pressure compressor powered by a five-stage low-pressure turbine. The low-pressure turbine operating at 3600 rpm drives a directly coupled 60 Hz generator. Generally, smaller gas turbines operate at high speeds and require a gearbox of different speed ratios to drive either a 60 Hz or 50 Hz generator. Because of the high-pressure ratio the exhaust temperature of the aeroderivative unit is with 450°C (842°F) relatively low. The relationship of pressure ratio to exhaust temperature is illustrated in Figure 5-21. A higher pressure ratio moves the compressor discharge temperature in the temperature/entropy diagram from point B to B'. At a constant inlet temperature the turbine inlet point moves from C to C' and the gas turbine exhaust temperature is reduced by the move of the turbine expansion line endpoint from D to D'. Although the increased pressure ratio results in an improvement of the simple-cycle efficiency, the combined-cycle efficiency does not improve. As discussed in the next chapter there is an optimal pressure ratio for achieving the best possible combined-cycle efficiency.

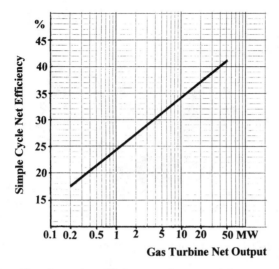

FIGURE 5-22. Simple-cycle efficiency of gas turbines smaller 50 MW.

The 41% gas turbine power plant net efficiency of the 42 MW aeroderivative engines is the most efficient simple-cycle arrangement. With decreasing unit ratings, the efficiency level is also lower as illustrated in Figure 5-22. The trend of the net efficiency level of smaller size gas turbines operating as simple-cycle units reveals that gas turbines of about 1 MW rating will reach only a roughly 24% simple-cycle efficiency level with the pressure ratio being reduced to about 5:1.

Like mid-size gas turbines, small gas turbines can be of either heavy-duty or aeroderivative design featuring multiple-stage axial compressor and turbine blade path arrangements. However, there are small gas turbine designs that combine axial and radial blading. An example is illustrated in Figure 5-23. The aeroderivative unit shown is a modified helicopter engine. The rating of this gas turbine for power generation is 3.0 MW at a speed of 15,000 rpm. With a pressure ratio of about 11:1, a simple-cycle plant net efficiency of 28.3% is achieved. The dual-shaft design features axial- and one-centrifugal compressor stages, and a turbine section with four axial stages.[34]

Figure 5-24 illustrates a 525-kW mini gas turbine as still a multiple-stage design with a two-stage centrifugal compressor and a three-stage axial turbine section. This gas turbine provides a power plant net effi-

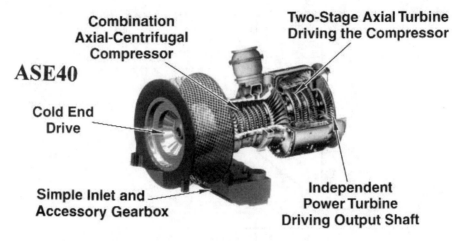

FIGURE 5-23. Aeroderivative 3.0 MW small gas turbine.

ciency as a simple-cycle unit of 21.3%. These small and mini gas turbines are ideally suited for co-generation.[34]

Smaller mini- and micro-turbines are generally of a single-stage centrifugal compressor and single-stage radial turbine design. Such low rated micro gas turbines achieve a power plant efficiency in the roughly

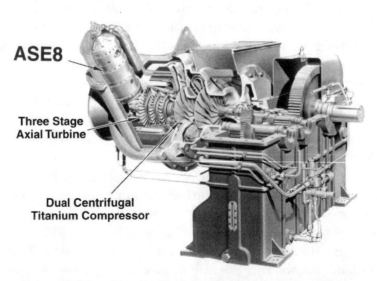

FIGURE 5-24. Aeroderivative 525 kW mini gas turbine.

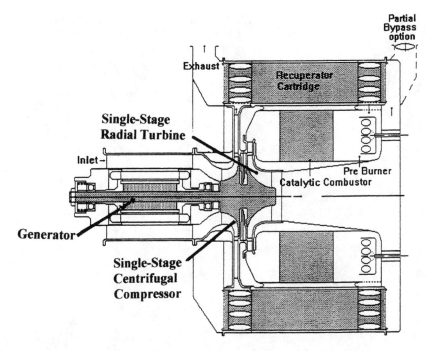

FIGURE 5-25. 5 kW micro gas turbine with recuperator.

10% range as simple-cycle units. However, because of their small size, they can be equipped with integrated recuperators to improve their efficiency. An example of a 5-kW micro gas turbine with a recuperator is illustrated in Figure 5-25. This micro gas turbine with recuperator can achieve a power plant net efficiency of about 20% with a pressure ratio of only 3:1 and at a speed of 150,000 rpm.[58] In the recuperator, the gas turbine exhaust gases with about 680°C (1250°F) temperature are utilized to heat the compressor discharge air to about 620°C (1150°F) before entering the combustion system. The exhaust gases leaving the recuperator maintain a temperature of roughly 230°C (450°F). This exhaust heat can still be used to improve the fuel utilization by adopting, for example, heat pumps or absorption chillers.

Another potential future application of mini and micro gas turbines can be their integration with high-temperature fuel cell systems. For increasing the electric power and electric efficiency, mini and micro

FIGURE 5-26. Micro steam turbine-generator in 1928.

combined-cycle power plants could be built. Micro steam turbines for power generation were already built in the early 1900s. Figure 5-26 shows a 1928 photograph of an 85 MW large steam turbine and a micro steam turbine-generator.[2] Such micro steam turbine-generators were built within the 0.5 kW to 5.0 kW rating range to provide lighting on ships and railroad trains.

CHAPTER 6

Combined-Cycle Power Plants

Combined-cycle power plants are built around gas turbines being the major prime mover generating roughly two-thirds of the plant output. The variety of combined-cycle power plant concepts is endless. Here we will concentrate on combined-cycle power plants with large full-size, heavy-duty gas turbines and heat recovery steam generators only. Combined-cycle plants can also be designed with any size gas turbine as prime movers.

The Department of Energy (DOE) established the goal for the power generation industry to build combined-cycle power plants with large heavy-duty gas turbines that will achieve a net efficiency of 60% and a single-digit NO_X emission without utilizing catalytic converters. This goal can only be achieved with heavy-duty gas turbines of the most advanced design, keeping their performance within a combined-cycle power plant in mind. Designing a gas turbine for the best performance in a combined-cycle power plant requires optimizing its pressure ratio, since the highest pressure ratio does not necessarily lead to the best combined-cycle power plant performance. Figure 6-1 illustrates in two diagrams the simple-cycle and combined-cycle power plant performance for gas turbines with increased firing temperature and pressure ratio.[59] The diagrams show the power plant efficiency over the specific output being the plant output per hot gas intake into the turbine in kg/s (lb/s).

In the first diagram the trend is revealed that with a simple-cycle application an increase in pressure ratio leads to a rise in plant efficiency, and an increase in firing temperature results in a rise in the specific plant output. Mid-size aeroderivative engines built as peakers in simple-cycle power plant arrangements reach an efficiency of 40% and more because of their high pressure ratio in the 30:1 range. The results for combined-cycle power plants illustrated in the second diagram show quite a different trend. Raising the firing temperature improves the performance in regard to both the plant efficiency and specific plant output. However, raising the pressure ratio reduces the specific output of a combined cycle power plant and provides an optimum in power plant efficiency at a relatively low pressure ratio, for example 16:1, for a firing temperature of 1280°C (2340°F). Since the optimal pressure ratio for combined-cycle power plants is relatively low, advanced heavy-duty gas turbines were originally designed for a pressure ratio of up to 17:1. This provides a relatively good efficiency for operation in simple-cycle applications as well

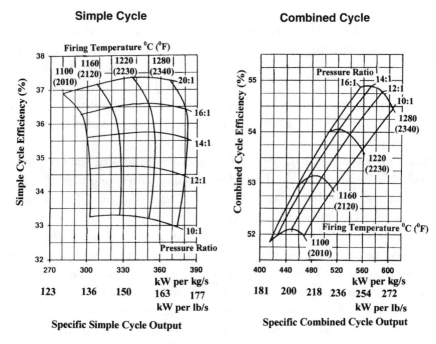

FIGURE 6-1. Performance improvement of simple and combined cycle.

as an optimal efficiency for combined-cycle power plants with a firing temperature in the 1300°C (2370°F) range.

In the meantime, the performance of advanced gas turbines has been further improved and the firing temperature has been raised to 1350°C (2460°F) and can reach a level of 1430°C (2600°F) in the near future. At this higher temperature level the optimal pressure ratio of combined-cycle plants will rise above 20:1 and combined-cycle plant efficiency of 60% can be expected.[60]

The major combined-cycle power plant concepts with different numbers of gas turbines are illustrated in Figure 6-2. The approximate size of a combined-cycle power plant is dependent on mainly the gas turbine rating and the number of gas turbine/heat recovery steam generator (HRSG) units that provide steam to the steam turbine. Typical examples of a one-to-one, a two-to-one, and a three-to-one gas turbine to steam turbine arrangement are shown. Assuming a gas turbine rating of about

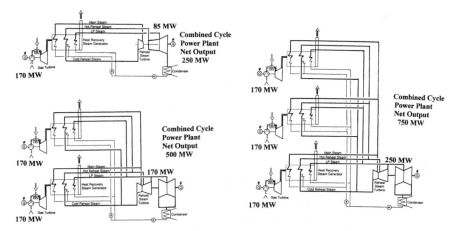

FIGURE 6-2. Combined-cycle power plants with one, two and three gas turbines.

170 MW, these three combined-cycle power plant concepts will generate approximately 250 MW, 500 MW, and 750 MW, respectively. This trend can continue and a four-to-one combined-cycle plant would be a 1000 MW power plant with a roughly 340 MW steam turbine.

The combined-cycle power plant concept combines a gas turbine Brayton process with a steam turbine Rankine process. This can be ideally done because advanced gas turbines discharge their exhaust gases at a temperature level of around 600°C (1100°F) and advanced steam turbines operate at about 565°C (1050°F) main and reheat steam. Heat recovery steam generators (HRSGs) have been developed and built to generate this steam with the gas turbine exhaust gases. Figure 6-3 illustrates a combined cycle with an advanced gas turbine Brayton process and steam turbine Rankine process. The advanced gas turbine cycle could be used in a peaking power plant to generate electric power from natural gas with a net power plant efficiency of about 38%. Combining this process with a triple-pressure reheat steam turbine cycle leads to a combined-cycle net efficiency of about 58%. The triple-pressure reheat steam cycle is highly efficient because it optimally utilizes the gas turbine exhaust heat to generate three steam flows in the HRSG as main steam, reheat steam, and low-pressure steam.

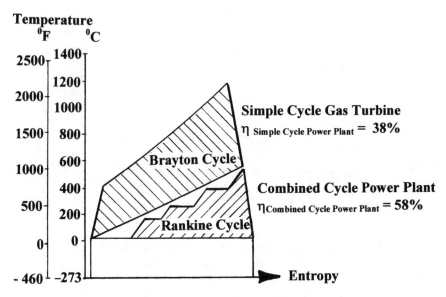

FIGURE 6-3. Combined-cycle power plant with advanced gas turbine and triple-pressure single-reheat steam cycle.

Other steam cycles can be applied but their combined-cycle power plant net efficiency will be lower. Figure 6-4 gives a comparison of the performance of different steam cycles, starting with a single-pressure non-reheat cycle as the basis.[61] The first step is a dual-pressure cycle, which increases the combined-cycle net efficiency by 3.5%. Selecting a dual-pressure reheat or a triple-pressure non-reheat cycle provides about the same 4.8% efficiency improvement over the single-pressure non-reheat base cycle. The triple-pressure reheat cycle as the best process provides a 6.2% combined-cycle power plant net efficiency improvement over the single-pressure non-reheat base cycle.

The triple-pressure reheat cycle mostly used with advanced heavy-duty gas turbines features an HRSG with a reheat section that not only reheats cold reheat steam from the steam turbine HP turbine section, but also generates additional hot reheat steam from the second pressure stage. The third steam supply is low-pressure steam that is superheated to a low temperature to be used as additional LP turbine inlet steam.

A steam turbine for such a combined-cycle power plant has to be designed for an increasing steam mass flow other than steam turbines for

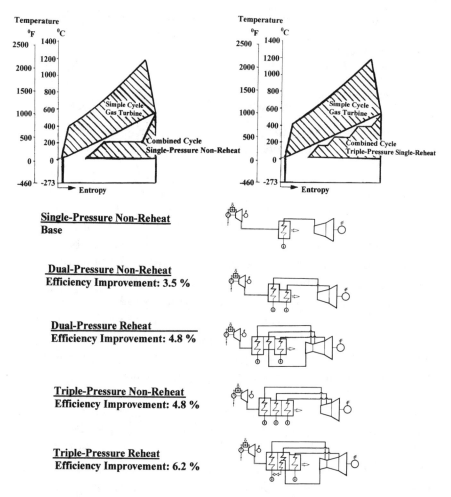

FIGURE 6-4. Steam cycles of combined-cycle power plants.

conventional power plants, which are designed for a decreasing steam mass flow because of their extractions for feedwater heating. Feedwater heating with extraction steam improves the performance of conventional steam plants because the low temperature boiler heat can be utilized for preheating the combustion air of the boiler instead of heating the feedwater in the boiler. With combined-cycle power plants the low temperature heat of the HRSG is most effectively used for heating the feedwater.

Steam Mass Flows	Main Steam	Reheat Steam	Exhaust Steam
Conventional Steam Turbine	100 %	94 %	65 %
Combined Cycle Steam Turbine	100 %	126 %	158 %

FIGURE 6-5. Mass flow comparison of conventional versus combined-cycle steam turbines.

Figure 6-5 compares the typical steam mass flows of a conventional reheat steam turbine and a triple-pressure reheat turbine of a combined-cycle power plant. If the main steam flow is assumed to be 100% for both types of turbine designs, then the exhaust steam flow of the conventional steam turbine is 65% of its main steam flow. However, the exhaust flow of the combined-cycle steam turbine is 158% of its main steam flow. Because the size of a steam turbine's largest section, namely the LP turbine section has to be designed for the exhaust steam flow, a combined-cycle steam turbine with the same backpressure as a conventional steam turbine features a much larger annulus area.

Optimizing the cold end of the combined-cycle by selecting the proper backpressure and last-stage blade length is important to achieve the best power plant performance. The annulus area, defined by the last-stage blade length, the rotor hub diameter, and the number of LP turbine flows, determines together with the volumetric exhaust steam flow the exhaust losses. Figure 6-6 shows exhaust loss curves of a variety of last-stage blades built into a single- and double-flow LP turbine. The example shown of a 100 MW combined-cycle steam turbine with a backpressure of 0.085 bar (2.5 in Hg abs) reveals that five different LP turbine configurations could be selected with exhaust losses from 17.4 kJ/kg to 59.3 kJ/kg (7.5 Btu/lb to 25.5 Btu/lb). The proper choice would most probably be the single-flow 32-in. LP turbine with an exhaust loss of 25.6 kJ/kg (11.0 Btu/lb).

When assuming that the optimal exhaust losses for a triple-pressure reheat combined-cycle power plant is 30 kJ/kg (12.9 Btu/lb), then the backpressure can be selected for each specific last-stage blade length and number of LP turbine flows. This is depicted in Figure 6-7 over the steam

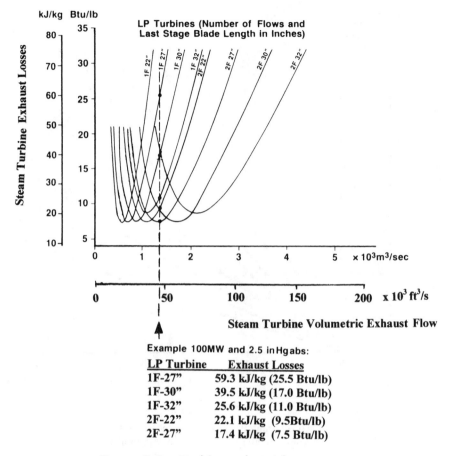

FIGURE 6-6. Turbine exhaust loss curves.

turbine output. As an example, a 170 MW-size steam turbine of a combined-cycle power plant with two advanced gas turbines could feature a single or double-flow LP turbine with last-stage blade lengths ranging from 660 mm (26 in.) to 965 mm (38 in.). For each specific LP turbine selection the backpressure is different, ranging from about 0.051 bar (1.5 in. Hg abs) to 0.27 bar (7.8 in. Hg abs). This backpressure range covers about 75 Btu/lb enthalpy drop or a difference in steam turbine output of about 20 MW or 12%, which relates to a 4% combined-cycle power plant efficiency. This reveals the importance of a low steam turbine backpressure to achieve a high combined-cycle efficiency.[62,63]

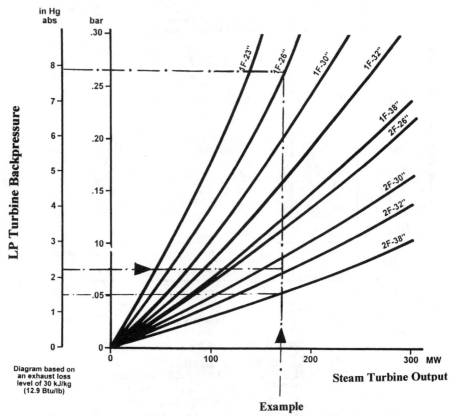

FIGURE 6-7. Selection of LP turbine size for triple-pressure reheat combined-cycle power plants.

For each specific site the backpressure depends on the available cooling water supply. In general three cooling systems could be selected, namely fresh water, wet cooling tower, or dry cooling system. As illustrated in Figure 6-8, the achievable backpressure becomes a function of the coolant temperature, and the initial temperature difference. The example shown is based on a wet cooling tower application with a coolant water temperature of 15°C (59°F) and an initial temperature difference of 25 K (45° R) providing a backpressure of 0.075 bar (2.2 in. Hg abs).[64]

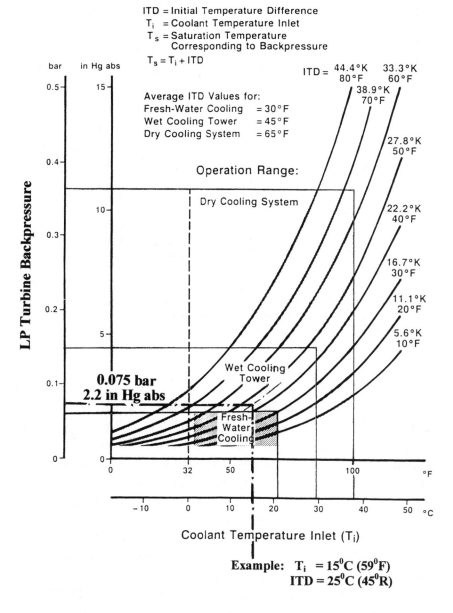

Figure 6-8. Selection of LP turbine backpressure for different cooling system site conditions.

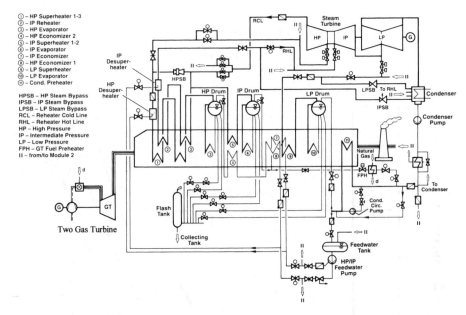

FIGURE 6-9. Triple-pressure reheat two gas turbine one steam turbine combined-cycle power plant concept.

Utilizing these site conditions in selecting the proper LP turbine size in accordance with Figure 6-7 leads to a double-flow LP turbine with 762 mm (30 in.) last-stage blades or a double-flow LP turbine with 813 mm (32 in.) last-stage blades for the 170 MW steam turbine.

A typical power plant arrangement with two advanced gas turbines and one combined-cycle steam turbine is shown in Figure 6-9 as a two-to-one arrangement.[65] The cycle shown is a triple-pressure reheat cycle. Two equally sized gas turbine/HRSG units provide main steam, reheat steam, and LP steam to one reheat steam turbine. The HRSGs feature an HP, IP, and LP drum followed by superheating sections to generate main and reheat steam at about 565°C (1050°F) and low pressure steam at about 260°C (500°F). The condensate is preheated in the HRSGs with low-temperature flue gas before the flue gases are discharged through the stacks at a discharge temperature of less than 100°C (212°F) when burning natural gas.

The natural gas is preheated with warm water from the IP economizer. Preheating the natural gas before combustion in the gas turbine has two

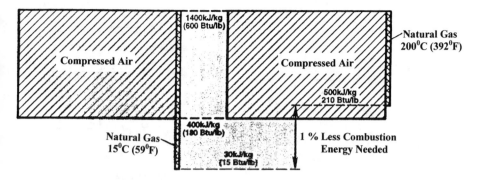

- Less Fuel Consumption...1 %
- Loss of Gas Turbine Output...............................0.1 %
- Loss of Steam Cycle Output:
  With Constant Stack Temperature...................1 %
  With 4°C (7.8°F) Lower Stack Temperature......0.55 %

LOWERING OF HEAT RATE BY 0.75 %

FIGURE 6-10. Preheating of natural gas fuel.

advantages: (1) it eliminates the potential of liquid hydrocarbons being present in the fuel, and (2) it enhances the combined-cycle power plant efficiency. The presence of liquid hydrocarbons must be avoided because of the potential of flashbacks in the gas turbine combustion system. The improvement of the combined-cycle efficiency by natural gas preheating is explained in Figure 6-10. The cycle improvement is due to the fact that low quality heat at the end of the HRSG is used to warm up the natural gas instead of having to heat the natural gas at the highest heat quality level in the combustion system. Since the natural gas is not heated by combustion heat energy from 15°C (59°F) to 200°C (392°F), about 1% of fuel can be saved. However, the gas turbine output is reduced by 0.1%, which accounts for 0.067% combined-cycle output. The steam turbine output is lowered by 0.55%, which accounts for 0.182% combined-cycle output. The total reduction of heat rate can then be calculated as 1% − (0.067 + 0.182) = 0.75%.

The steam turbine, which receives the steam supply from the two gas turbine/HRSG units, is a two-casing turbine with a combined HP/IP turbine section and a double-flow LP turbine section as illustrated in Figure 6-11. Main and reheat steam are admitted into the combined HP/IP

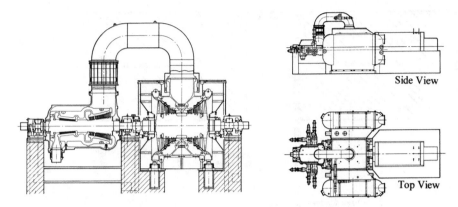

Figure 6-11. Two casing double-flow reheat system turbine with side condensers.

turbine and the low-pressure steam is added to the LP turbine inlet steam through the cross over piping. The double-flow LP turbine can be built with different last-stage blade sizes to provide an optimal annular area for each specific application. For combined-cycle power plants the LP turbine is designed with side exhausts for arranging two 50% condensers on either side of the turbine. This compact arrangement can be set on a low foundation slab avoiding a high-bay foundation.

HP, IP, and LP bypass systems are provided for best operating flexibility. For control and matching of the main and reheat steam temperatures de-superheating stations are installed. The operating flexibility of this two-to-one power plant concept is depicted in Figure 6-12. Independent operation of each gas turbine with its HRSG is performed with the isolation valves in the main, cold reheat, hot reheat, and LP steam lines being properly opened or closed. Main steam can be bypassed to the cold reheat line, cooling the reheater and being discharged from the hot reheat line into the condenser. The main bypass station's de-superheating station reduces the main steam temperature to the cold reheat steam temperature level. After reheat, the hot reheat steam is discharged into the condenser with the reheat steam bypass station, reducing the steam temperature to a level acceptable for the condenser. For flexible operation of the HRSG's LP section, the LP steam bypass station leads the LP steam after de-superheating also into the condenser. In this mode of operation, a single gas turbine can be started up to match the condi-

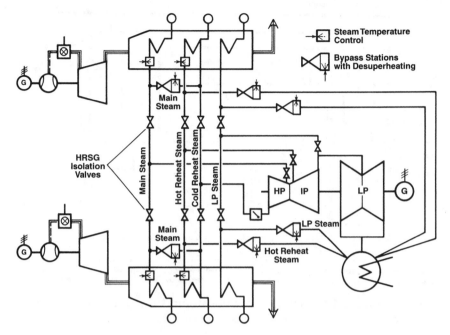

FIGURE 6-12. Operating flexibility of two gas turbine one steam combined-cycle power plant.

tion of the steam turbine, which can be either in operation with the other gas turbine/HRSG unit or on turning gear. After conditions are matched, switchover from bypass operation to steam turbine operation can be performed by opening the isolation valves and closing the bypass stations. With this arrangement, load operation with one or two gas turbines can be performed in a wide steam turbine loading range from 100% to 25% or less. Similar cycles are used for one-to-one and three-to-one gas-to-steam turbine power plants; however, one-to-one arrangements with relatively small steam turbines can utilize simpler cycles, e.g., a dual-pressure non-reheat cycle with a simpler HRSG design and a non-reheat steam turbine.

For advanced gas turbines in a one-to-one arrangement single-shaft gas/steam turbine units have been developed, as illustrated by an example shown in Figure 6-13. For best performance a triple-pressure reheat cycle was selected. The advanced 180 MW gas turbine is solidly coupled to the common 300 MVA size hydrogen-cooled generator on one side.

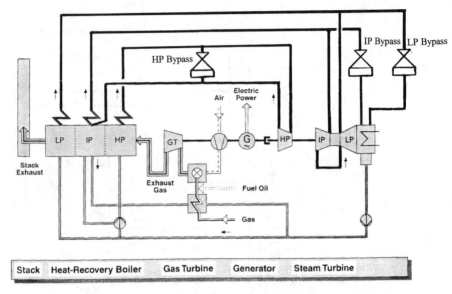

FIGURE 6-13. Single-shaft gas/steam turbine combined-cycle design concept.

For the small steam turbine rating of 85 MW, a steam turbine with an HP turbine section and a combined IP/LP turbine section was selected. The 85 MW steam turbine is connected to the other generator rotor end via a synchronous clutch. This arrangement provides excellent operating flexibility. In conjunction with the HP, IP, and LP steam bypass systems, the gas turbine and HRSG can be quickly started without the steam turbine because the generated stream from the HRSG is bypassed into the condenser. As soon as the generated steam matches the requirements for steam turbine startup, the steam turbine speed is raised from turning gear speed to rated speed, the synchronous clutch automatically engages, the steam turbine output is raised, and the steam bypass systems closed.

The steam turbines for combined-cycle power plants have been designed based on the experience with conventional power plant turbines. The basic configurations of combined-cycle steam turbines are illustrated in Figure 6-14. Non-reheat and reheat steam turbines of different sizes are available as single-casing and two-casing steam turbines. A variety of LP turbines with different last-stage blade length are available. The

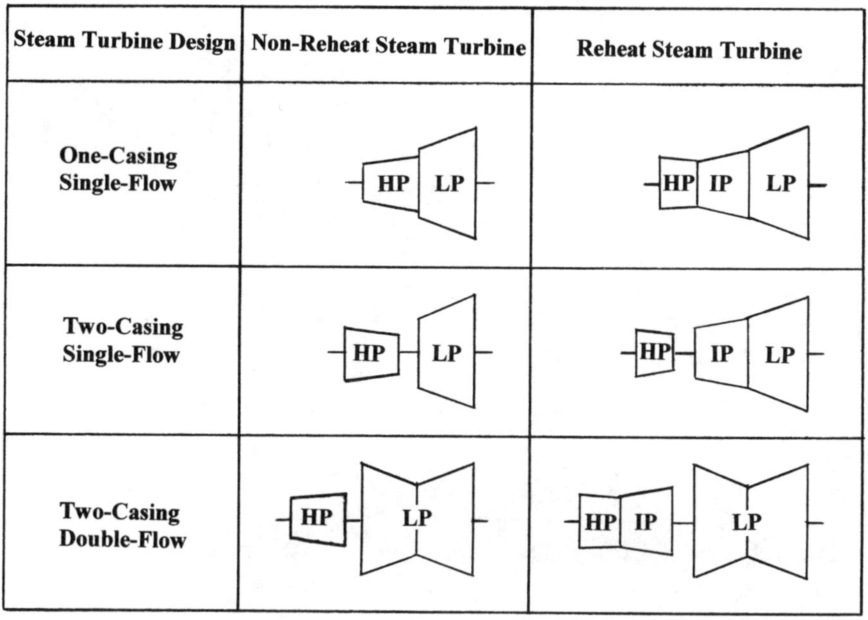

Steam Turbine Design	Non-Reheat Steam Turbine	Reheat Steam Turbine
One-Casing Single-Flow	HP | LP	HP | IP | LP
Two-Casing Single-Flow	HP — LP	HP — IP | LP
Two-Casing Double-Flow	HP — LP	HP | IP — LP

FIGURE 6-14. Steam turbines for combined-cycle power plants.

LP turbine exhaust can be designed for single-flow units axial or downward, and for double-flow units sideways or downward, depending on the selected power plant arrangement. The axial as well as the sideways arrangements are used to put the turbines on a low foundation slab. Figure 6-15 illustrates such an arrangement as a single-shaft gas/steam turbine unit with the common generator in the center, an advanced gas turbine on the left, and a two-casing single-flow triple-pressure reheat steam turbine on the right. The axial exhaust and condenser arrangement minimizes the exhaust losses and allows for a gas/steam turbine unit with a centerline height of only 5.2 m (17 ft) above the foundation slab.

The introduction of advanced gas turbines with higher exhaust temperatures of about 650°C (1200°F) leads to main and reheat steam temperatures in excess of 565°C (1050°F). Building compact turbines for combined-cycle power plants requires the design of steam turbine rotors, which are exposed to high steam temperatures on one end and low-temperature operating conditions at the other end. These operating conditions require high-creep rupture strength on one end of the rotor and

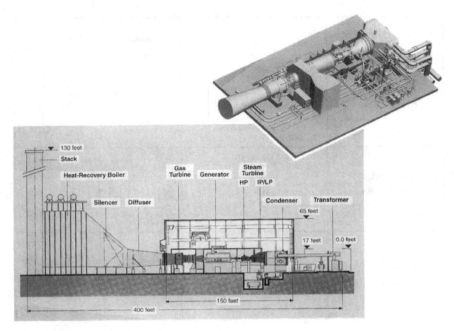

Figure 6-15. Single-shaft gas/steam turbine arrangement.

high-fracture toughness on the other end. Single-casing reheat steam turbines with advanced steam conditions require specially heat-treated forgings or rotors consisting of two different steel alloy forgings being welded together.[66]

Because combined-cycle steam turbines have large exhaust flows, large LP turbine sections are needed. To build a large annulus area LP turbine, larger last-stage blades were developed. Last-stage blades up to 965 mm (38 in.) in blade length from steel forgings and 1070 mm (42 in.) titanium blades are available for 60 Hz applications. For 50 Hz applications these blades can be scaled-up to become 1390 mm (55 in.) and 1540 mm (61 in.) last-stage blades, respectively.

Figure 6-16 shows a reheat steam turbine with a welded IP/LP turbine rotor and titanium last-stage blades of 1070 mm (42 in.) blade length for 60 Hz application. The two-casing turbine features a single-flow HP turbine and a single-flow combined IP/LP turbine with an axial exhaust section. This turbine is specifically designed for single-shaft gas/steam turbine arrangements with the generator drive at the HP turbine end

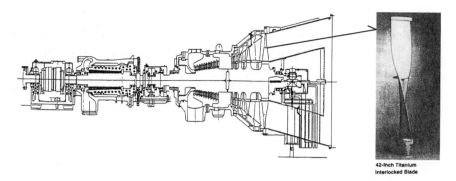

FIGURE 6-16. Two-casing single-flow reheat steam turbine.

through the shown synchronous clutch. The welded IP/LP turbine rotor consists of a high-creep strength chromium/molybdenum/vanadium (CrMoV) forging and a nickel/chromium/molybdenum (NiCrMo) forging for low-temperature operation with high-fracture toughness. To build the large annular area single-flow LP turbine section the 1070 mm (42 in.) last-stage blade is made from titanium to minimize the blade's weight and, consequently, the blade's centrifugal forces. The yield strength of the titanium blade material with 930 Mpa (135 ksi) is as high as the yield strength of a steel alloy blade material. However, the specific weight of titanium with 4430 kg/m^3 (0.16 lb/in.^{3}) is only 57% of the 7750 kg/m^3 (0.28 lb/in.^{3}) specific weight of steel alloys for blades. As shown, the titanium blades are of an interlocked design with integral shrouds.

The output of combined-cycle plants can be increased by power augmentation. The gas turbine output can be raised by injecting water or steam into the combustion system or by fogging or evaporative cooling systems installed upstream of the compressor inlet. The steam turbine output can be quite drastically increased with duct burners being installed into the HRSG. However, such a burner system provides additional output at a reduced combined-cycle power plant efficiency level. Today, combined-cycle power plants are built with a net power plant efficiency of around 58% and the heavy-duty, full-size gas turbine technology has evolved to an advanced level that allows for 60% efficiency combined-cycle power plants burning natural gas. With this performance they are also the cleanest fossil-fueled power plants. As previously shown in Figure 2-34, their specific carbon dioxide (CO_2) discharge of 340 g/kWh

(0.73 lb/kWh) is about one-half of the CO_2 discharge of an advanced coal-fired power plant and about one-fourth of a 1970s vintage lignite-fueled plant. There is only a minute discharge of sulfur dioxide when burning natural gas, and also the specific NO_X discharge per kilowatt-hour of an advanced gas turbine combined-cycle power plant is about 60% of that of a coal-fired power plant equipped with a DENOX system.

CHAPTER 7

Repowering Steam Turbine Plants With Gas Turbines

In the previous chapter we discussed only one kind of combined-cycle plant, namely plants in which steam for the steam turbine is generated in heat recovery steam generators (HRSGs) only. Before discussing repowering it is important to realize that other combined-cycle power plant concepts have been developed and utilized for new and repowering applications. Figure 7-1 illustrates the three basic concepts of repowering steam turbine plants with gas turbines to combined-cycle power plants. The example shown is based on the use of a 150 MW gas turbine with a simple-cycle efficiency of about 36% and an exhaust temperature of about 550°C (1020°F) for the three repowering concepts.[67]

At first, with the basic gas turbine, HRSG, and steam turbine cycle (HRSG repowering), an existing average 75 MW steam turbine power plant can be repowered with one 150 MW gas turbine to generate approximately 150 + 75 = 225 MW output at an efficiency level of about 54%. The second concept is the fully fired repowering also referred to as topping or hot windbox repowering. Here the gas turbine exhaust gases are fed into the rebuilt existing boiler as preheated combustion air. The optimal arrangement can achieve an efficiency of about 49% when repowering a 450 MW size existing reheat steam turbine plant to become an approximately 450 + 150 = 600 MW repowered plant. The third repowering concept utilizing the gas turbine exhaust heat for feedwater heating is referred to as feedwater heater repowering or parallel repowering. The latter term is used when not only feedwater heating is performed, but also some additional steam is generated. This concept of repowering can be applied to any size of existing steam plants, including nuclear power plants. The efficiency of the repowered plant depends greatly on the steam plants rating or, better yet, the ratio of gas to steam turbine output of the repowered unit. The two examples shown are of a 750 MW and a 200 MW reheat steam turbine plant achieving, after repowering with a 150 MW gas turbine, a power plant efficiency of 46% and 50%, respectively. The efficiency increases with smaller steam turbine power plant ratings because relatively more feedwater is preheated by the gas turbine exhaust gases, reducing the need for feedwater heating by steam turbine extractions toward zero.

The application range of any repowering concept can possibly be found in the mid-range power demand, as illustrated in Figure 7-2.[68] For peaking demand, simple-cycle gas turbines have been installed because of

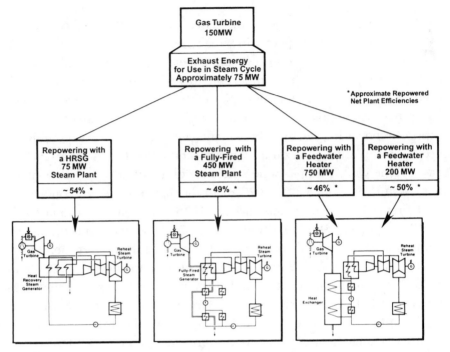

FIGURE 7-1. Basic repowering concepts with a 150 MW gas turbine.

their low capital costs, simplicity, and operating flexibility. With few full-load operating hours there is no need for highly efficient power generation. Their application range is below 1000 full-load operating hours per year. The diagram shows the relative fuel consumption, the efficiency, and the specific cost difference as a function of full-load operating hours per year. In the low operating hour range, the simple-cycle gas turbine is the right choice because low fuel cost and the fuel cost difference between a simple-cycle and a combined-cycle does not justify investing in more costly power plant options. On the other hand for base-load operation in the full-load operating range between 7000 hours and 8760 hours per year, the green field combined-cycle plant seems to be the right choice. The specific cost difference of roughly $200/kW compared to a simple-cycle peaking gas turbine power plant can be justified and is compensated by the large difference in relative fuel consumption per year.

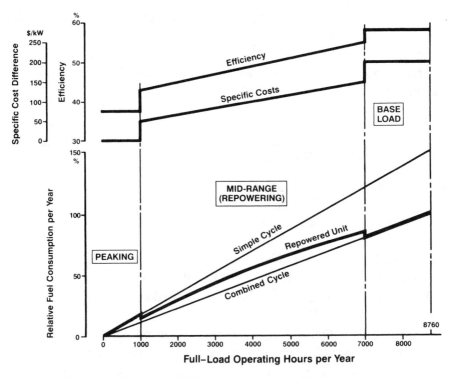

FIGURE 7-2. Potential repowering application range.

However, for the mid-range power generation repowering could become the most desired choice. Figure 7-2 shows potential repowering options with increasing specific cost differences. Feedwater heating and parallel repowering are certainly options that can be built very cost effectively with only a small specific cost increase over a peaking gas turbine plant. Repowering with gas turbine/HRSG units is the more expensive alternative with a higher specific cost difference. As shown in the diagram, they fit best into the higher full-load operating hour range, since the latest HRSG repowering projects with advanced gas turbine/HRSG units have reached efficiency levels equal to green field combined-cycle power plants. Fully fired repowering has been used in the past, but not lately because of the extensive modifications required on existing boilers. In addition, advanced gas turbines do not have a sufficiently high oxygen content in their exhaust gases to be used for fully fired repowering. A repowering project should be started by identifying the power generation need and by performing a condition assessment of the existing steam turbine power plant. Depending on the expected full-load operating hours of the repowered unit, either high efficiency or low investment costs become the priority.

HRSG REPOWERING

Standard combined-cycle power plants and HRSG repowering projects have to be developed around the selected gas turbine and its performance data. On the other hand an existing steam turbine power plant should be utilized as much as possible with a minimum amount of costly modifications. To match these two demands a wide variety of HRSG repowering options have been developed. The following examples should provide sufficient information to select the optimal repowering concept for each specific application.

As illustrated in Figure 7-3, a major portion of the existing power plant will be eliminated and replaced with a gas turbine/HRSG unit. The steam turbine and condenser can be utilized, and by properly designing the HRSG only minor modifications of these systems are needed. An optimal fit is a gas turbine rating of about twice the rating of the steam turbine. The example shows a 75 MW non-reheat steam turbine to be repowered with a 150 MW size gas turbine, utilizing a dual-pressure HRSG to provide main and LP steam.[69]

The next example is a case of a relatively old 140 MW reheat steam turbine plant to become a low mid-range unit with highest operating flexibility. Because there was a concern about the remaining life of the IP turbine rotor, it is recommended that the reheat steam turbine be converted into a non-reheat steam turbine and that a dual-pressure non-reheat HRSG be utilized, as depicted in Figure 7-4.[70] This relatively low-cost concept also eliminates the entire reheat steam pipe system. The IP turbine is only exposed to very low steam temperatures, which eliminates the concern about excessive low-cycle fatigue even though the unit will be exposed to frequent starts. To avoid high moisture content in the LP turbine section a throttle valve arrangement is provided. This valve arrangement also controls excessive stressing of the HP turbine last-stage blading. This change in operating mode can be performed without any

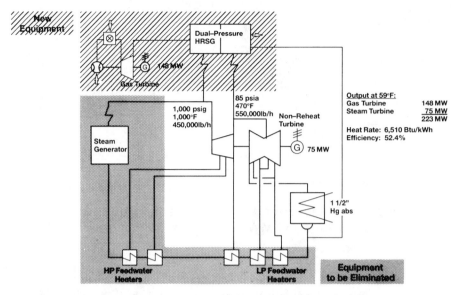

FIGURE 7-3. HRSG repowering of a 75 MW non-reheat steam turbine power plant with a 150 MW gas turbine.

changes in the steam turbine's blade path, making this repowering a low-cost approach.

Since a 150 MW gas turbine was considered, the gas-to-steam turbine ratio was too small and additional steam turbine capacity could be utilized by adding duct firing in the HRSG. With this arrangement the most efficient performance is achieved at about 220 MW net power plant output with a net power plant efficiency of about 52%. With duct firing the steam turbine output can be increased from 72 MW to 120 MW, resulting in a maximum power plant net output of 267 MW at a reduced plant net efficiency level of 47%. This is still much higher than the power plant net efficiency of only 38% before repowering. Ideally, the additional output can be used as peaking capacity provided at a very low investment cost.

Repowering large reheat steam turbines with gas turbine/HRSG units requires installation of more than one unit. A 300 MW reheat steam turbine needs about 600 MW gas turbine capacity or 4 × 150 MW gas turbines. Such an arrangement is illustrated in Figure 7-5. To provide high operating flexibility the arrangement shown of four gas turbines allows

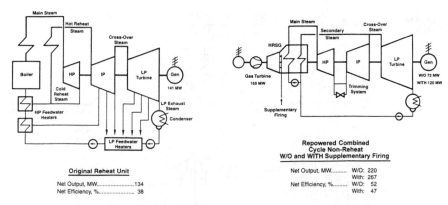

FIGURE 7-4. Repowering a 140 MW reheat steam turbine power plant with a 150 MW gas turbine to a non-reheat cycle.

for the operation of the steam turbine with any of the gas turbine/HRSG units. The dual-pressure reheat cycle is equipped with main steam, cold reheat steam, hot reheat steam, LP bypass steam, and feedwater headers. Each gas turbine/HRSG unit has shut-off valves and an HP bypass system to bypass main steam into the cold reheat system. With this arrangement the steam turbine can be in operation with any number of gas turbines, and an additional one can be started with its HRSG bypassing main steam through its HP bypass and the common LP bypass into the condenser. After the steam conditions are matched in bypass operation the steam can be supplied into the common headers and from there to the steam turbine. This repowering concept is ideally suited for load cycling with a possible operating range from 900 MW full load with four gas turbines down to less than 100 MW with one gas turbine in operation. This entire operating range can be covered at an efficiency level from 55% to 45%, by switching gas turbines on and off as shown in the performance diagram. When operating with all four gas turbines at part load, the efficiency would drop to 45% already at about 400 MW.

The next example is quite a challenging repowering project that asks for the repowering of a 660 MW reheat steam turbine power plant with three advanced 50 Hz gas turbines.[71] The goal is to achieve close to green field combined-cycle power plant performance when in operation with the gas turbines and still keep all the equipment to also allow operation with the existing boiler burning a different fuel. Figure 7-6 shows the ex-

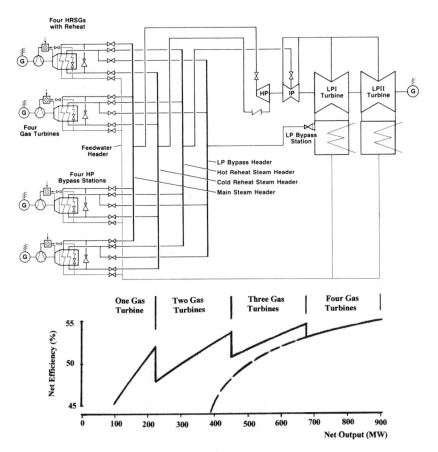

FIGURE 7-5. Repowering a 300MW reheat steam turbine power plant with four 150 MW gas turbines.

isting reheat steam cycle with a main steam pressure of 160 bar (2300 psig) and a main and reheat steam temperature of 538°C (1000°F). The existing 660 MW power plant provides a plant net efficiency of about 39%. When operating the existing steam turbine with three 270 MW gas turbines, a total output of 1210 MW can be generated with a 57% plant net efficiency. At this operation mode the steam turbine output is 400 MW. To operate the steam turbine in the most efficient way the main steam pressure is reduced to 75 bar (1080 psig) to match the volumetric flow conditions of the original steam path design. Since the mass flow conditions are different for the operation with the boiler and 660 MW versus the operation with the three gas turbine/HRSG units and 400 MW,

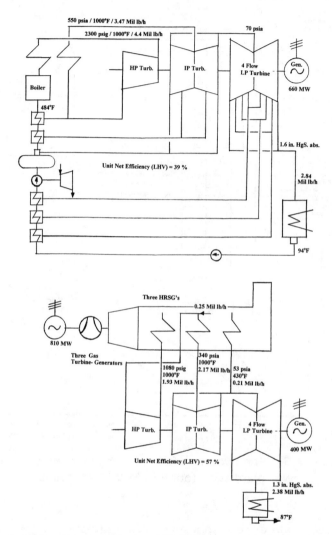

FIGURE 7-6. Repowering a 660 MW steam turbine plant with three advanced 50 hertz gas turbines.

the repowered cycle has lower pressure levels, listed in Figure 7-7 as post-repowering values versus pre-repowering values. The post-repowering pressure levels allow the design of an almost optimal combined-cycle triple-pressure reheat combined cycle without any modifications of the steam turbine.

Figure 7-8 illustrates how the two different power plant cycles are connected to each other and how they can be operated separately. A com-

Steam Turbine Pressure		Repowering	
		Pre	Post
Main Steam	psig	2300	1080
	bar	160	75
Reheat Steam	psia	550	340
	bar	38	23
LP Steam	psia	70	53
	bar	4.8	3.7
Back Pressure in. HgS. abs.		1.6	1.3
	bar	0.054	0.044

FIGURE 7-7. Steam turbine pressure levels at pre- and post-repowering conditions.

bination of shut-off valves and bypass systems allows independent start-up of the boiler, as well as each gas turbine/HRSG unit. This repowering concept has the ability to operate in a hybrid mode with both the boiler and the HRSGs supplying steam to the steam turbine. Such an operational mode is shown in Figure 7-9, with the three gas turbine/

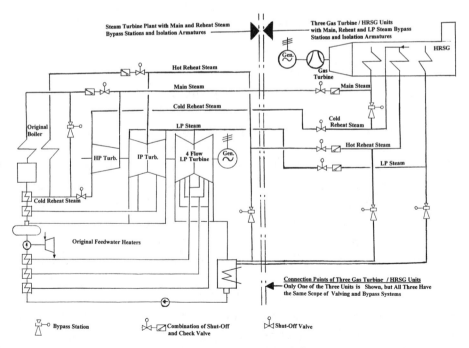

FIGURE 7-8. Repowering concept for maximum operating flexibility.

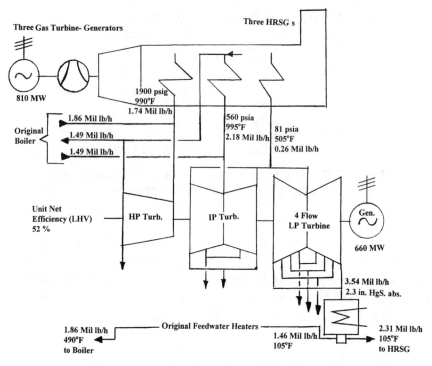

FIGURE 7-9. Operation in a hybrid mode with boiler and HRSGs.

HRSGs at full load and the boiler supplying enough steam to increase the steam turbine output to the original 660 MW level. However, this mode of operation results in higher mass flows and pressure levels in the LP turbine than for which it was originally designed. Under these operating conditions the total output is 1270 MW, which can be generated at a power plant net efficiency of 52%. This operating flexibility also provides fuel flexibility because electric power can be generated by burning the original fuel in the boiler and by burning natural gas or #2 fuel oil in the gas turbine combustion systems. However, not only the heat rate, but the emissions will be much lower when operating the advanced gas turbines with natural gas.

The advantage of repowering with natural gas fueled gas turbines instead of operating a 1970 vintage lignite-fueled power plant is illustrated in Figure 7-10.[72] The reduction of carbon dioxide (CO_2) discharge for a repowering project of a 675 MW lignite-fueled power plant with three

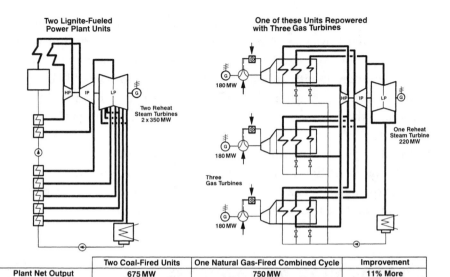

	Two Coal-Fired Units	One Natural Gas-Fired Combined Cycle	Improvement
Plant Net Output	675 MW	750 MW	11% More
Net Heat Rate (LHV)	11000 Btu/kWh (31%)	6100 Btu/kWh (56%)	45% Better
CO_2 Discharge	16.8 x 10⁹ lb/Yr	5.1 x 10⁹ lb/Yr	70% Less
Reduction in CO_2 Discharge is 5 850 000 Short Tons Per Year			

FIGURE 7-10. Reduction of CO_2 discharge by HRSG repowering.

180 MW advanced gas turbines is revealed. One of the two 350 MW steam turbines is repowered to become a 220 MW steam turbine receiving steam from three gas turbine/HRSG units. The second steam turbine is shut down and could be used for later repowering.

The power plant output post repowering has been increased by 11% and the net power plant heat rate (efficiency) has been improved by 45%. The CO_2 discharge is reduced by 70%. The repowered unit discharges only 30% CO_2 of the original total power plant even though the output has been increased by 75 MW from 675 MW to 750 MW. The total CO_2 discharge has been reduced by 5.85 million short tons (5.3 million metric tons) per year. If both the original steam turbines would be repowered, the power plant output would double to 1500 MW and still only 60% of the original CO_2 would be discharged.

The 5.85 million short tons (5.3 million metric tons) of CO_2 reduction can be related to the total CO_2 discharge of all U.S. power plants. Presently, all U.S. power plants discharge about 2 billion short tons (1.8 billion metric tons) per year into the atmosphere. Without any effort to reduce CO_2 discharge from power plants, a yearly increase of 25 million short tons (22.7 million metric tons) has been forecast. If only one lignite-fueled power station of 675 MW would be repowered with advanced gas turbines, the 5.85 million short tons (5.3 million metric tons) of CO_2 reduction with an increase of 75 MW output would be a significant benefit. Five of such repowering projects per year in the U.S. would eliminate the forecasted increase of 25 million short tons (22.7 million metric tons).

In addition, a switch to natural gas fuel would practically eliminate the discharge of SO_2 and could minimize the NO_X emission by a factor of approximately 17, assuming the original coal-fired power plant was not equipped with catalytic converters (see Figure 2-33).

HOT WINDBOX REPOWERING

Hot windbox or fully fired repowering utilizes the gas turbine exhaust as hot air for combustion in the existing boiler. This combined-cycle concept was introduced in the early 1960s.[73, 74] Gas turbines of the 1960 vintage had about 15% oxygen in their discharge gas and were well suited for combustion in modified boilers. A large number of such power plants were built with a wide variety of different fuels from coal to natural gas being burned in the boiler. They were originally built as hot windbox units or were repowered by adding a gas turbine and modifying the boiler at a later date. A typical example of repowering an existing 590 MW reheat steam turbine power plant with a 140 MW, 50 Hz gas turbine is illustrated in Figure 7-11. Besides the gas turbine, an auxiliary stack, a small FD-fan, a furnace bypass as well as HP and IP partial flow economizers were added.

With the repowered station the output of 690 MW could be reached and the full load net power plant efficiency was increased from 40.2% to 46.6%. The post-repowered plant efficiency curve reveals a very flat shape in a wide partial load range, since full load is reached with the auxiliary fan in operation to provide combustion air for the boiler in parallel to the gas turbine.

The gas turbine and the boiler of this station burn natural gas. Besides the 100 MW more output and the efficiency improvement of 11.6%, the NO_X emission was drastically improved by operating the gas turbine in

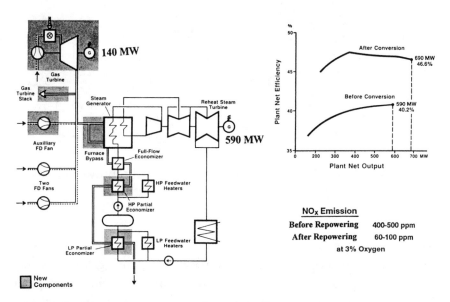

FIGURE 7-11. Hot windbox repowering of a 590 MW reheat steam plant.

a premix mode to achieve a single-digit dry low-NO_X emission level when burning natural gas in the gas turbine based on a 15% oxygen content. The NO_X emission of the entire unit (gas turbine and boiler) was reduced from a 400 ppm to 500 ppm range to a level of 60 ppm to 100 ppm at 3% oxygen content.

Even though hot windbox repowering was successful, it has not been considered lately because rebuilding an old power plant is very costly. In addition, the efficiency level that can be achieved is not as high as that of a HRSG repowering and the advanced gas turbines are not well suited for hot windbox repowering because their exhaust gas oxygen level has dropped below 13%.

FEEDWATER HEATER AND PARALLEL REPOWERING

The feedwater heater repowering was introduced to provide an improved performance over the performance of a peaking gas turbine installa-

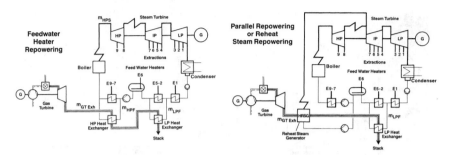

FIGURE 7-12. Feedwater heater and parallel repowering.

tion.[69, 74] With feedwater heating, additional power is generated in an existing steam turbine because feedwater heating with gas turbine exhaust heat reduces or eliminates the extraction flow to the steam plant feedwater heaters and increases the steam flow through the steam turbine resulting in an increase in its output. Figure 7-12 shows such an arrangement with HP and LP heat exchangers to preheat the feedwater utilizing the gas turbine exhaust energy. Since the exhaust temperature of the gas turbine is high enough to generate steam, the feedwater heater repowering can also be modified to become a parallel repowering by generating some reheat steam in parallel with the existing boiler. This parallel repowering provides better performance than the feedwater heater repowering, but needs a single-pressure HRSG combined with the LP feedwater heater.

The different results of feedwater heater versus parallel repowering are shown in Figure 7-13 for the repowering of reheat steam power plants with a range of pre-repowering power plant efficiencies from 38.9% to 42.3%. The difference is the result of an assumed reheat steam turbine efficiency range of 10%. Repowering is performed with an advanced midsize 70 MW gas turbine that can achieve a simple-cycle efficiency of 36.5% and a combined-cycle efficiency of 54% at a gas turbine to steam turbine ratio of about 2.

The total power plant efficiency can be increased at about the same magnitude with both the feedwater and parallel repowering methods up to a gas turbine output to steam turbine output ratio of 0.17 or 17%. If the ratio is large, the feedwater heater repowering does not bring a significant improvement and even drops above a gas turbine to steam tur-

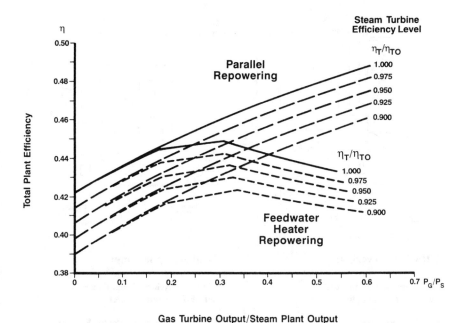

FIGURE 7-13. Total plant efficiency for feedwater heater and parallel repowering of a steam turbine plant with different original efficiency levels.

bine ratio of 30%. The parallel repowering achieves a steady improvement in total power plant efficiency even beyond the shown 0.6 or 60% gas turbine to steam turbine output ratio. Based on this result both repowering concepts have their application range and an optimization for a specific application should be performed to select the best repowering concept.

The optimal combined-cycle power plant is a gas turbine/HRSG/steam turbine arrangement with the gas turbine output being roughly twice the output of the steam turbine, as shown previously in Figure 6-2 for 170 MW gas turbine applications. Optimal HRSG repowering should be based on this approximate 2:1 gas/steam turbine ratio. Such HRSG repowering provides about 200% more power plant output. However, a smaller increase is often achieved because the steam turbine has been derated to better match the repowered steam cycle conditions. An increase in excess of the 200% would require a steam turbine uprating, which could involve major component changes.

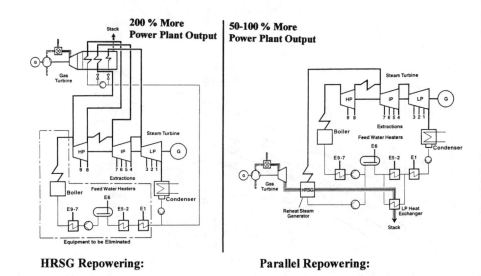

FIGURE 7-14. Full HRSG repowering versus parallel repowering.

If a much smaller power plant output increase is needed, parallel repowering should be applied. As illustrated in Figure 7-14, the power plant output increase of 50% to 100% can be covered instead of the about 200% of HRSG repowering. For a 170 MW steam turbine power plant repowering, a 60 MW gas turbine would provide about 85 MW or 50% more power plant output. The gas turbine to steam turbine output ratio is 60 MW/195 MW ≙ 31%. Raising the steam turbine output from 170 MW to 195 MW, or by less than 15%, is generally possible without major steam turbine modifications.

A 115 MW gas turbine would provide for the same 170 MW steam turbine power plant, roughly a 170 MW or 100% output increase, and the gas turbine to steam turbine output ratio after repowering is 115 MW/225 MW ≙ 51%. The steam turbine output has to be raised from 170 MW to 225 MW or by 32%, which requires major steam turbine modifications.

Repowering the same 170 MW steam turbine power plant with a smaller gas turbine, e.g., a 30 MW gas turbine, would lead to a gas tur-

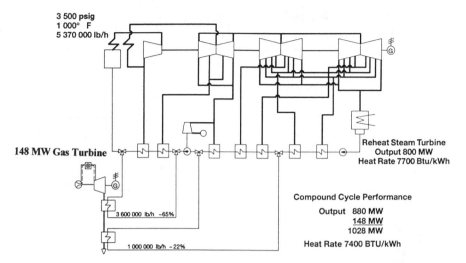

FIGURE 7-15. Feedwater heater repowering of 800 MW steam turbine power plant.

bine to steam turbine output ratio of 30MW/180 MW ≜ 16.7% and should possibly be best performed by feedwater heater repowering. Such a low gas turbine to steam turbine output ratio repowering example is illustrated in Figure 7-15, where an 800 MW steam turbine power plant is repowered with a 148 MW gas turbine. HP and LP partial flow feedwater heaters preheat about 65% of the high-pressure feedwater and 22% of the low-pressure feedwater. The result is 228 MW increase in power plant output and 300 Btu/kWh or 3.9% improved heat rate. The gas turbine to steam turbine output ratio after repowering is very low with only 148 MW/880 MW ≜ 16.8%.

In case the selected gas turbine is too large for feedwater heater repowering or for parallel repowering of one steam turbine in a multiple unit station, the feedwater heater repowering of two steam turbines could be considered. As illustrated in Figure 7-16, repowering two 300 MW steam turbine power plant units with one 150 MW gas turbine increases the net power plant output from 585 MW to about 800 MW. The steam turbine output of each unit was increased from 300 MW to 333 MW or by only 11%, providing a total steam turbine output increase of 2 × 33 MW = 66 MW. The overall gas turbine to steam turbine output ratio is 150 MW/666 MW ≜ 22.5%.

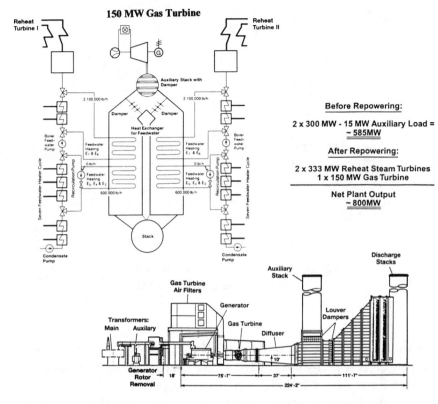

FIGURE 7-16. Feedwater heater repowering of 2 × 300 MW steam turbine units with 1 × 150 MW gas turbine.

Feedwater heater and parallel repowering has the advantage of providing some fuel flexibility. A parallel-powered station with a coal-fired boiler and a natural gas-fired gas turbine is illustrated in Figure 7-17, which generates 320 MW base load with the coal-fired boiler and an additional 100 MW when running the 70 MW gas turbine with natural gas. The gas turbine exhaust heat is utilized in a HRSG, which generates 540°C (1004°F) reheat steam to be fed into the reheat steam turbine. In addition, the HP feedwater and the LP feedwater are preheated in parallel to the steam turbine extraction feedwater heaters. The power station is designed to also provide pressure-controlled extraction steam from the steam turbine for co-generation and features modern emission controls.

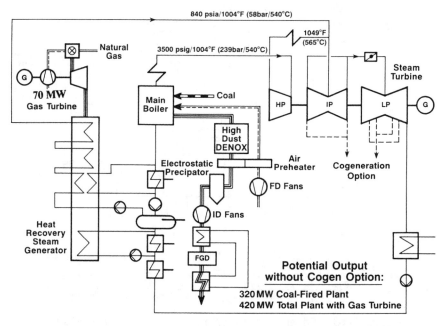

FIGURE 7-17. Parallel powered coal-fired steam plant with natural gas-fired gas turbine.

Since feedwater and parallel repowering still operates the existing boiler, the improvement in emission is based on the ratio of gas turbine to boiler combustion. The example shown in Figure 7-18 has the steam turbine derated from 400 MW to 300 MW and a 180 MW advanced gas turbine added. The total power plant output increased by 80 MW and the net power plant efficiency improved from 31% to 39%. This parallel repowering concept features a HRSG design, which generates reheat steam and preheats LP feedwater in an LP heat exchanger section. Despite the 20% increase in power output the CO_2 discharge has been reduced by 25%.

The reduction of all emissions depends highly on the ratio of gas turbine to boiler fuel consumption or the ratio of pre- to post-repowering output. Figure 7-19 reveals the CO_2 discharge reduction for the repowering of lignite-fueled 300 MW to 500 MW power plants with one advanced 180 MW gas turbine. Depending on the derating of the steam turbine, the CO_2 discharge reduction can be as high as 55% when adding

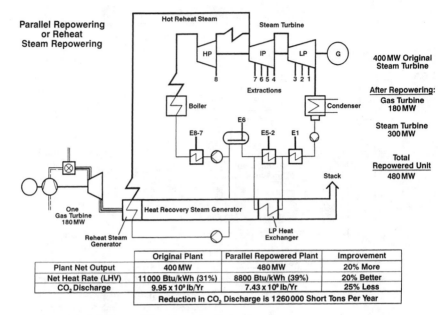

FIGURE 7-18. Parallel repowering of 400 MW lignite-fueled station with a 180 MW advanced gas turbine.

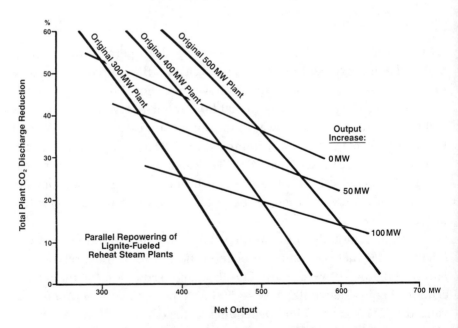

FIGURE 7-19. CO_2 reduction by parallel repowering with one 180 MW gas turbine.

no power plant output, and as low as 10% when increasing the power plant output by 100 MW.

Several repowering projects with gas turbines being added to existing steam turbine power plants have been successful. Their success was related to providing more power at a higher efficiency level from existing power plants at strategic locations in the grid system and to the reduction of mainly NO_X and SO_2 emissions. Little or no credit was given to the fact that such repowering of old steam turbine power plants drastically reduces the CO_2 discharge. If such credit will be considered in the future, repowering will become even more attractive.

Based on experience the three repowering concepts:

- feedwater heater repowering
- parallel repowering
- HRSG repowering

should be applied depending on the gas turbine to steam turbine output ratio. Any steam turbine power plant can be repowered by selecting the most economical solution. Steam turbines can be operated at quite different steam conditions very efficiently without major modifications if those steam condition changes have been properly established.

CHAPTER 8

Coal Gasification and Fuel Cell Combined-Cycle Plants

Integrating a coal gasification facility with a combined gas/steam turbine cycle power plant forms a very efficient coal-fired power generation plant concept, whereas the integration of fuel cells with a combined gas/steam turbine cycle plant provides the most efficient natural gas-fired power generation station concept.

The development of coal gasification combined-cycle power plants started about 30 years ago with the gasification and the combined-cycle power plant technologies as separate, already proven, technologies. Relatively small capacity fuel cells are available today. Building larger facilities and combining them with combined gas/steam turbine cycles could lead to natural gas-fired power plants with efficiencies greater than 70%.

INTEGRATED COAL GASIFICATION COMBINED-CYCLE POWER PLANTS

Integrated coal gasification combined-cycle (IGCC) power plants have been developed to build the most efficient coal-fired power plants with low emission levels. The first coal gasification combined-cycle power plant went into operation in 1972 at the Luenen power station in Germany.[75] It featured five fixed bed gasifiers, a 74 MW gas turbine, and a 96 MW non-reheat steam turbine. A unique feature of this pilot plant is the two pressurized steam generators directly mounted to the gas turbine, replacing the two silo-type combustion chambers of the 1960 vintage gas turbine. The steam generator arrangement and the plant's cycle concept are illustrated in Figure 8-1. The plant net efficiency was 37% based on the low heat value (LHV) of the coal. The pressurized steam generator operated within about a 10 bar (150 psia) pressure level; a booster compressor raised the pressure for the steam and air-blown gasifiers to a 20 bar (300 psia) level. The gas turbine inlet temperature was relatively low at 820°C (1508°F); the main steam for the steam turbine was rated at 525°C (977°F). The pressurized steam generators burned coal gas as its main fuel, but also had burners installed to operate with fuel oil as a secondary fuel. Another coal-fired power plant concept, namely combined-cycle power plants with pressurized fluidized bed combustion systems, has also been developed and successfully operated.

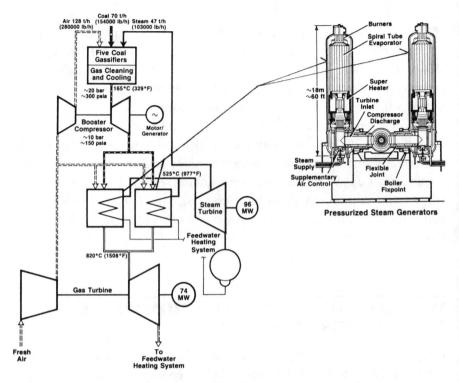

FIGURE 8-1. Early coal gasification plant Luenen, Germany.

The next IGCC pilot plant was the Cool Water project in California, featuring an oxygen-blown gasifier and an 80 MW gas turbine. The plant net output was about 120 MW. Presently, about 30 large integrated coal gasification combined-cycle (IGCC) power plants have been put into operation worldwide. However, some of these power plants are burning either refinery residues or orimulsion instead of coal. There are a large variety of gasification suppliers, providing three different types of coal gasifiers, as illustrated in Figure 8-2: fixed bed, fluidized bed, and entrained bed units.[76]

Fixed bed gasifiers generally use lumped coal. The coal enters from the top and falls in counter-flow to the rising gasifying agent consisting of steam and oxygen. It is first dried, then pyrolyzed, and finally gasified. The average reaction temperature is in the range from 800°C to 1000°C (1470°F to 1830°F). In the upper part of the gasifier, during preheating and pyrolysis, the coal releases tars and oils. These substances

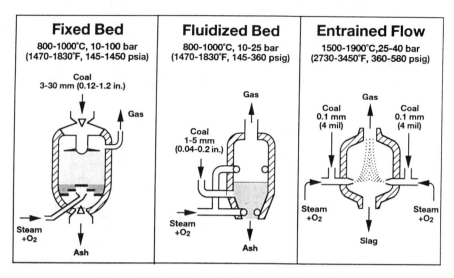

FIGURE 8-2. Three basic types of gasifiers.

condense when cooling the raw gas. A reliable waste recovery is not possible due to the fact that the coal gas is quenched by water injection. This operation simultaneously removes dust and liquid hydrocarbons.

The fluidized bed gasifiers use fine grained coal, which is converted to COH_2 in the fluidized bed by means of the gasifying agent's oxygen and steam. The gasification temperature is restricted to the range below the softening point of the ash at 800°C to 900°C (1470°F to 1650°F). Only with lignite and highly reactive hard coal can the process perform a relatively high carbon conversion in the 90% to 95% range. Therefore, the residual coke must be fed to a furnace, as with a fluidized bed combustor.

In the entrained flow gasifiers fine grained coal is converted to a tar and oil-free crude gas at a high temperature level of 1500°C to 1900°C (2730°F to 3450°F), using either oxygen or enriched air. The coal can be fed dry or as a coal/water slurry. The carbon conversion is nearly 100% and ash is removed as nonleachable slag.

The three different types of gasifiers also require somewhat different IGCC power plant concepts, as illustrated in Figure 8-3. Compared to the other gasifier concepts, the fixed bed gasification produces fuel with the highest heating value, which needs the lowest amount of oxygen, and

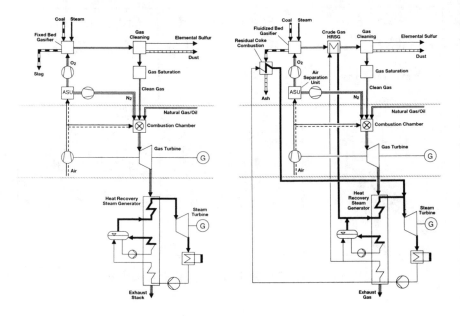
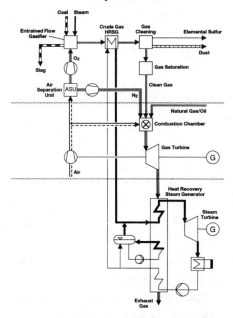

FIGURE 8-3. Integration of the three basic type gasifiers into combined-cycle plants (IGCC).

which can be equipped with only a small air separation unit. Consequentially, only a small amount of nitrogen is available for admixing to the coal gas and the final gas to be burned has a high heating value with a relatively high NO_X formation during combustion. To achieve low NO_X emissions, steam has to be added to the coal gas fuel. The amount of added steam injection could be restricted because of gas turbine operating limitations. The crude gas downstream of the gasifier is quenched with water. Therefore, no heat recovery in the form of steam generation is available. The effect on the overall plant concept is a low power output of the steam turbine. The integration is minimal because there is only a small amount of heat and mass flows between the gas generating unit and the power generation unit.

Considering the heating value of coal gas, oxygen consumption, and raw gas heat recovery, the fluidized bed gasification concept lies between the fixed bed and the entrained flow gasification. With regard to steam production, a portion of the feedwater is extracted from the gas turbine's HRSG and fed to the crude gas HRSG. To limit the tube temperature in the crude gas HRSG, which is operating in a highly corrosive environment, only saturated steam is produced. Subsequently, this steam is superheated in the gas turbine's HRSG. If applicable, the residual coke combustion can provide an additional amount of main steam.

The entrained flow gasification produces raw gas with the lowest heating value at the highest oxygen consumption. Therefore, the largest air separation system is needed, which provides the highest nitrogen mass flow for NO_X depression. With this integrated plant concept, very low NO_X emissions can be achieved without any steam or water injection. Due to a high outlet temperature, the portion of sensible heat in the raw gas is high.

Consequently, the steam and heat flows between the gasification system and the power generation plant are the highest compared to the other two gasification systems. The entrained flow gasification with dry coal feed seems to provide the best overall net plant heat rate and efficiency if full integration is applied.

When designing a coal gasification power plant it is important to perform an economic evaluation of the justifiable integration (partial or full) and to identify the operational limitations of the considered gas turbine model. An important issue is the permissible imbalance of the gas flows

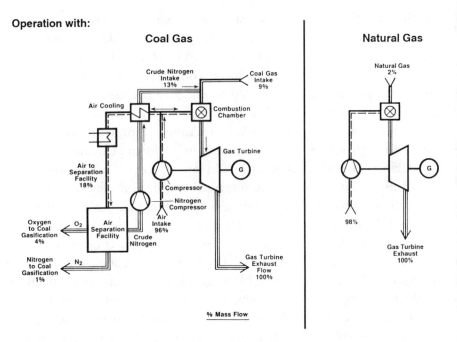

FIGURE 8-4. Integration of gas turbine with air separation facility.

through the compressor versus the turbine section of the gas turbine. Figure 8-4 shows how a full integration of the gas turbine with the air separation facility can balance these compressor and turbine flows. The basic gas turbine design considers natural gas as fuel. With natural gas as fuel, the 100% gas turbine flow consists of 98% air and 2% natural gas. With the full integration, the gas turbine compressor provides 96% air, from which 18% is fed to the air separation facility, providing 96% − 18% = 78% to the turbine section. The turbine section also receives 13% nitrogen and 9% coal gas flows, totaling 78% + 13% + 9% = 100%.

If partial integration is considered and the turbine flow becomes higher than the compressor flow, the gas turbine output increases significantly, because turbine power is directly used for power generation without the need for air compression. However, this might not be a gain in net power plant performance since a separate compressor is needed.

The flowchart of a fully integrated IGCC power plant is shown in Figure 8-5.[77] The gasification is built as an oxygen-blown entrained flow

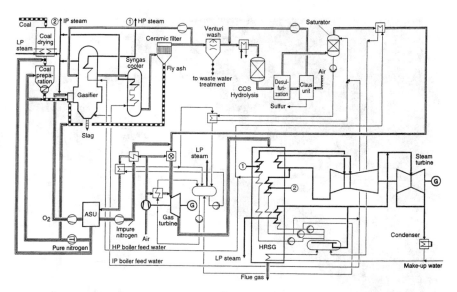

FIGURE 8-5. Fully integrated 300 MW IGCC power plant.

gasifier and the power plant features a reheat steam cycle. Steam is additionally generated in the coal gasification facility and fed into the steam cycle as main and reheat steam for the steam turbine, whereas LP steam from the steam turbine is used in the gasification facility. This highly integrated IGCC power plant was built and is generating about 300 MW net output at a net efficiency level of about 45%. Today, more advanced gas and steam turbine technology would raise this efficiency level.

Presently, IGCC power plants can be built to provide net power plant efficiencies below the 50% threshold, utilizing low-temperature gas clean-up systems. Further improvement in performance beyond the 50% mark and simpler plant concepts with minimized integration could be realized if hot gas clean-up systems were to become commercially available and provide the purity equivalent to that achieved with the low-temperature wet gas purification processes.

Experience with the combustion of low-Btu gases in gas turbines was gained in the early days of heavy-duty gas turbines. For combustion of coal gas, the early experience with burning blast furnace gas can be applied. When comparing the heat values based on the low heat value (LHV)

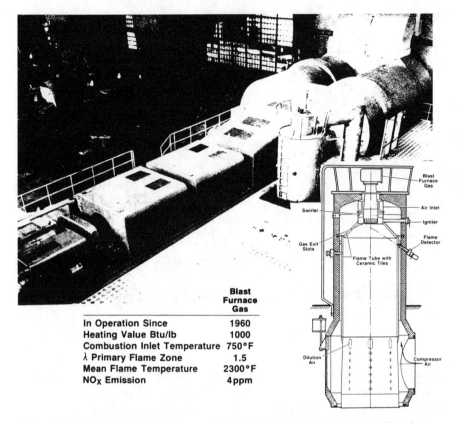

	Blast Furnace Gas
In Operation Since	1960
Heating Value Btu/lb	1000
Combustion Inlet Temperature	750°F
λ Primary Flame Zone	1.5
Mean Flame Temperature	2300°F
NO_X Emission	4 ppm

FIGURE 8-6. 1960-vintage gas turbine for low-Btu gas combustion.

of the different fuels, it can be seen that the heat values of the gas turbine fuels are quite different:

	kJ/kg	Btu/lb
Blast furnace gas	2,300	1,000
Air-blown coal gas	4,000–6,000	1,700–2,600
Oxygen-blown coal gas	8,000–11,000	3,400–4,700
Natural gas	50,000	21,500
#2 fuel oil	42,000	18,100

Low-Btu fuel has the advantage that it can form a very lean mixture with air to combust with a low NO_X emission. As an example, a blast furnace application from the early 1960s is illustrated in Figure 8-6. This 7 MW gas turbine features a silo-type combustion chamber with one

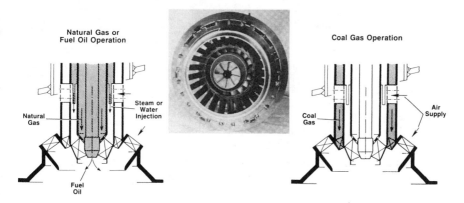

FIGURE 8-7. Triple fuel burner for IGCC power plants.

burner. The blast furnace gas is combusted at a mean flame temperature of 1250°C (2300°F) with a NO_X emission of only 4 ppm.[78]

However, the challenge of designing large gas turbines that burn low-Btu gas is the handling of the large amounts of fuel in a controlled manner. For a natural gas-fired unit the fuel flow is roughly 2% of the compressor air discharge flow. Since the heat value of coal gas can be less than one-tenth of the heat value of natural gas, as much as 20% fuel has to be mixed with the combustion air.

In addition, large amounts of steam and/or nitrogen have to be mixed to the coal gas to reduce the NO_X emission of IGCC plants to less than 25 ppm. These systems require large control and stop valves as well as large-diameter piping. The burners have to be designed to handle not only the coal gas fuel, but are often required to also handle natural gas and/or oil as secondary fuels. Figure 8-7 shows a burner built for combusting coal gas, natural gas, or #2 fuel oil. The triple fuel burner and its controls have been designed for fast switch-over from one fuel to another if a fuel supply is interrupted.

Coal gasification combined-cycle (IGCC) power plants operate successfully and produce a more efficient performance than pulverized coal-fired power plants. A comparison of an advanced 42% efficient pulverized coal-fired power plant with a 46% efficient fully integrated IGCC power plant is shown in Figure 8-8. Because of its increased efficiency, the IGCC plant consumes roughly 10% less coal and discharges about 10% less CO_2 when generating the same 600 MW net output. The pul-

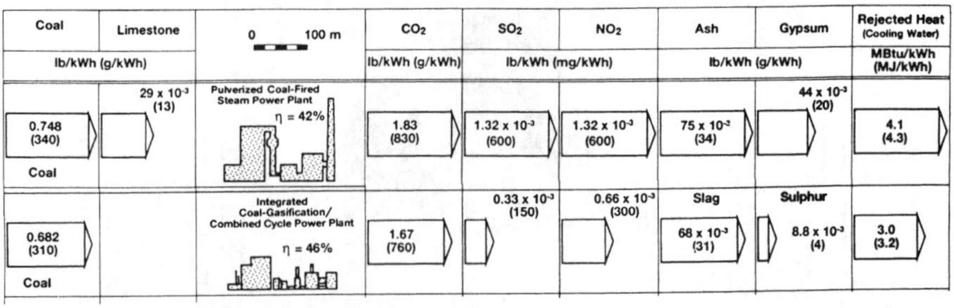

166 • 100 YEARS OF POWER PLANT DEVELOPMENT

Coal	Limestone	0 100 m	CO_2	SO_2	NO_2	Ash	Gypsum	Rejected Heat (Cooling Water)
lb/kWh (g/kWh)			lb/kWh (g/kWh)	lb/kWh (mg/kWh)	lb/kWh (g/kWh)			MBtu/kWh (MJ/kWh)
0.748 (340) Coal	29 x 10^{-3} (13)	Pulverized Coal-Fired Steam Power Plant $\eta = 42\%$	1.83 (830)	1.32 x 10^{-3} (600)	1.32 x 10^{-3} (600)	75 x 10^{-3} (34)	44 x 10^{-3} (20)	4.1 (4.3)
0.682 (310) Coal		Integrated Coal-Gasification/ Combined Cycle Power Plant $\eta = 46\%$	1.67 (760)	0.33 x 10^{-3} (150)	0.66 x 10^{-3} (300)	Slag 68 x 10^{-3} (31)	Sulphur 8.8 x 10^{-3} (4)	3.0 (3.2)

FIGURE 8-8. 600 MW coal-fired power plant supply flows, emissions, and by-products.

verized coal plant needs limestone for the wet scrubber. The SO_2 and the NO_X emissions of the IGCC plant versus the pulverized coal plant are about one-fourth and one-half, respectively. Pulverized coal plants discharge ash and gypsum, while IGCC plants discharge slag and sulfur. It's important to note that these discharge products can be used as by-products of the power generation process. The slag and the relatively small amount of sulfur become available as marketable by-products in IGCC plants. With the combined cycle of the IGCC plant, only a portion of the power generation is performed in the steam turbine. Therefore, the steam turbine exhaust flow is small, and only a small amount of exhaust steam has to be condensed. The total heat rejection of the IGCC plant is only 73% of that rejected by the pulverized coal plant.

FUEL CELL COMBINED-CYCLE POWER PLANTS

Fuel cells as a source of electricity have been under development in the last four decades. They are currently utilized mainly in the defense industry and in aerospace applications. Pilot facilities have been built to gain experience with fuel cells in the power generation industry. Low and high temperature fuel cell types have been developed, such as the polymer electrolyte membrane fuel cell (PEMFC), which operates at a temperature level below 140°C (284°F), and the solid oxide fuel cell (SOFC), which operates at up to 1000°C (1830°F). The later developed solid oxide fuel cells are particularly promising for the power generation

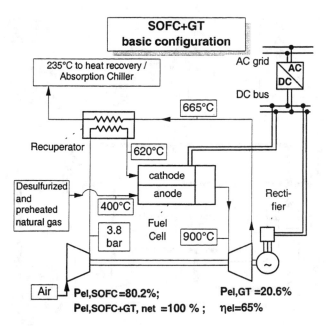

FIGURE 8-9. Solid oxide fuel cell and gas turbine power plant.

industry. Their simple-cycle efficiency is about 50% and the high operating temperature makes them well suited for combined plant concepts in which the fuel cell heat discharge is recovered by a gas turbine cycle.

Such a power plant concept — with a solid oxide fuel cell (SOFC) and a micro gas turbine (GT) — is illustrated in Figure 8-9.[79] This plant concept is also called hybrid cycle or SOFC + GT cycle. The fuel cell receives compressed air at 3.8 bar (55 psia) and 620°C (1150°F) from the gas turbine compressor and the recuperator. Desulfurized and preheated natural gas is used as fuel for the SOFC, which generates 80.2% of the electric power and heats the discharge gas up to 900°C (1650°F). In the turbine section of the gas turbine, this SOFC discharge gas expands to 665°C (1230°F), driving the compressor section and generating 20.6% electric output. The 100% net output of such a SOFC + GT power plant can be generated with a net electrical efficiency of 65% and the plant still provides 235°C (455°F) discharge gas downstream of the recuperator, which can be recovered for heating and cooling. For an installation in a building complex, heat pumps and/or absorption chillers can be installed. Supplying warm water, heating, and cooling by utilizing the ex-

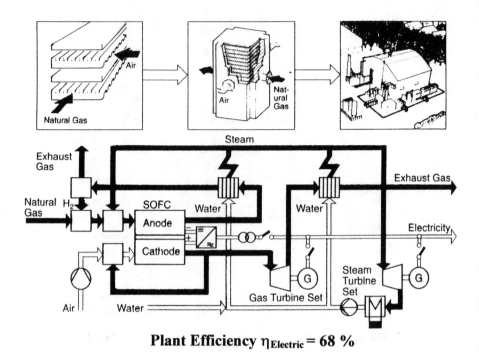

Plant Efficiency $\eta_{Electric} = 68\ \%$

FIGURE 8-10. Solid oxide fuel cell combined-cycle power plant. (Source: Siemens AG)

haust heat downstream from the recuperator provides a significant rise of the overall fuel utilization.

However, it must be realized that this promising technology is still in development. The major challenge is to build low-cost, high power density fuel cell assemblies, micro gas turbines with recuperators, etc., which operate reliably and sustain their high efficiency level. It is also important for the power generation industry to achieve operational flexibility, such as in load cycling, and frequent shutdowns and restarts.

An example of a plant for electric power generation only is illustrated in Figure 8-10. This power plant concept utilizes a gas turbine and a steam turbine. It features two heat recovery steam generators (HRSGs), one downstream of the SOFC and the other one downstream of the gas turbine. Both HRSGs generate main steam for the steam turbine. With this arrangement an electric net power plant efficiency of 68% has been estimated.

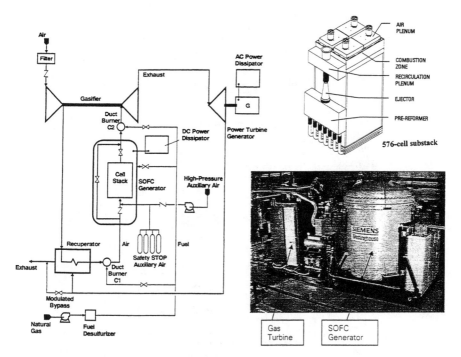

FIGURE 8-11. First 217 kW SOFC + GT pilot power plant.

Presently, the first hybrid SOFC + GT plant for 217 kW output is under construction with a 187 kW SOFC assembly and a 47 kW micro gas turbine. This pilot plant concept with its pressurized SOFC is illustrated in Figure 8-11 and features two of the shown 576-cell substacks.[80] The SOFC and gas turbine are skid mounted with the following approximate dimensions: 7.4 m (24.3 ft) length, 2.8 m (9.2 ft) width, and 3.9 m (12.8 ft) height. The electrical net efficiency of this first pilot plant has been estimated at 57%, plus the hot water or heat supply from the heat recovery system.

The possibility of future coal gasification and fuel cell technology allows us to build power plants that are a combination of both. Such a potential coal-fired fuel cell power plant concept is illustrated in Figure 8-12. An IGCC power plant provides fuel for solid oxide fuel cells and the heat from the SOFC is recovered in the gas turbine/steam turbine

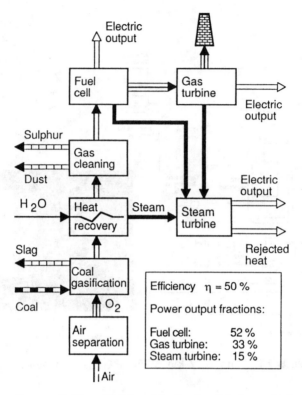

FIGURE 8-12. Solid oxide fuel cell/integrated coal gasification plant.

combined-cycle power plant. The fuel cells, burning cleaned coal gas, generate about 52% of the plant output and the gas and steam turbines together generate the remaining 48%. A combined fuel cell/IGCC power plant concept could raise the 45% efficiency level of present IGCC power plant concepts to a 50% net power plant efficiency level.[81]

CHAPTER 9

Future Power Generation

The future of the power generation industry will be shaped by the increasing demand for electricity, the availability and cost of fuel, and the development of new technologies, as well as their fuel consumption, capital and operating costs, and, last but not least, environmental considerations. Predictions have been made that electric power generation will increase from 12 to 23 trillion kWh per year between 1990 and 2020. The predicted fuel sources and their increased use are illustrated in Figure 9-1.[59] Coal-fired power plants will still be the major suppliers of electric power in the year 2020, generating 38% of the world's electricity. The generation of electricity with fuel oil will decrease to 8%. The number of power plants, mainly combined-cycle power plants burning natural gas, will grow drastically from 1990 to 2010, generating 17% instead of 12% of the world's electric power. After 2010 it is expected that this trend will slow down and natural gas-fired power plants in 2020 will generate even less, namely 15% of the world's electricity. Nuclear power generation will increase only slightly and lose 1% of its present 17% share. Hydropower generation as a renewable source of energy will increase by the year 2020 to reach the 20% level of the global electric power generation. The electricity supply from all other renewable sources will still be small even though an increase from 1% to 3% is expected, a 200% growth. The requirements for improving power plants using different fuels are also listed in Figure 9-1 and show that all power plant concepts should be built at reduced costs to be the most competitive. The present trend towards combined-cycle power plants is based on the fact that these plants are low cost and fast to install. In addition, the efficiency is high, emissions low, and low-price natural gas is available, at least in portions of the world.

However, the world's fuel sources are limited, as illustrated in Figure 9-2. The coal reserves account for 67%, while natural gas accounts for only 14.2% of the proven recoverable sources, which reveals that 4.7 times more coal than natural gas reserves is available.[1] Depending on the information source, predictions of the remaining fuel availability are widespread. Based on the 1992 Madrid World Energy Council, the worldwide availability (in number of years) of different fuels is as follows:

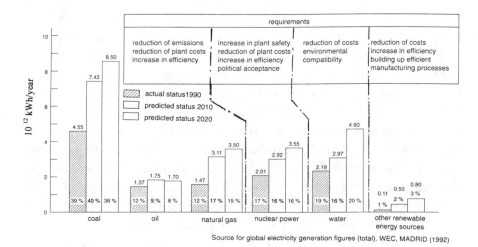

FIGURE 9-1. Predicted trend of global electric power generation from different fuel sources.

Coal	250 years
Fuel oil	45 years
Natural gas	60 years
Nuclear fuel (indicates fuel recycling)	3300 years

Not everyone will agree on these numbers because new fuel reserves will be found, but this can be assumed for these kinds of fuel. It is important to realize that coal seems to be available 4.2 times longer than natural gas and nuclear fuel will be available 13.2 times longer than coal. Fuel oil has the shortest availability period, and since it is used for needs other than electric power generation, in the future it should not be burned in power plants except the oil refinery residues in IGCC power plants.

Fuel availability in the long run will influence the power generation industry to a greater extent. Another fuel issue is its availability, or better non-availability, in different countries. The major portion of the world population, namely 76%, lives in developing countries. This 76% uses only 20% of the total electric power worldwide. By the year 2020 it is expected that this percentage of electricity consumption in developing countries will increase to 50%. Figure 9-3 reveals the imbalance of population to electric power consumption. The industrialized countries (including the Eastern European countries) have an average electricity con-

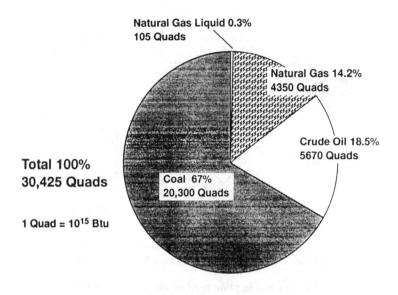

FIGURE 9-2. Proven worldwide recoverable fossil energy reserves as of the end of the 1990s.

sumption of 7200 kWh per person, whereas the developing countries' electric power consumption is only 600 kWh per person. The electric power consumption in the U.S. is 12,000 kWh per person and is 20 times higher than the average consumption in developing countries.

Countries such as China and India depend mostly on coal as fuel. Because of limited financial resources of developing countries with mainly coal resources, advanced clean pulverized coal or IGCC power plants are hardly affordable. The emissions of these countries will increase and, consequently, the industrialized countries must reduce their emissions to compensate for the increased emissions in developing countries.

The worldwide situation is best illustrated by the CO_2 discharge causing the greenhouse effect. According to the committee, "Protecting the Earth's Atmosphere," in 1990 about 22 billion metric tons per year (24.2 billion short tons per year) were discharged into the atmosphere by all countries worldwide. This equates to about 16 billion metric tons per year (17.6 billion short tons per year) from industrialized countries versus 6 billion metric tons per year (6.6 billion short tons per year) from

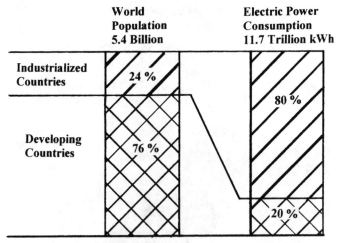

Sources: World Bank Development Report '91, BP Statistical Rev. of World Energy '91, IEA'95

Electric Power Consumption per Person in the Year 1990:
World Population 2 200 kWh
Industrialized Countries 7 200 kWh
Developing Countries 600 kWh

FIGURE 9-3. Imbalance between world population and electric power consumption in the year 1990.

developing countries. It has been recommended to reduce the total worldwide CO_2 discharge, and at the same time allow the developing countries to increase their portion. A more realistic approach would be to maintain the CO_2 discharge at a constant level by allowing developing countries to increase their discharge and requesting the industrialized countries to reduce their portion. It was predicted by the Protecting the Earth's Atmosphere Committee that by the year 2020 the CO_2 discharge from developing countries will increase to 8 billion metric tons per year. Keeping the total amount constant would mean that the industrial countries have to reduce their discharge to 14 billion metric tons per year, which relates to 2 billion metric tons per year over a 30-year span. This accounts for an annual CO_2 discharge reduction of 2000/30 = 67 million metric tons per year. Since the power generation industry discharges about 25% of the total CO_2 discharge, it could be expected that the power generation industry in the industrial countries will reduce their share of 25% from 67 million metric tons per year, namely 17 million metric tons

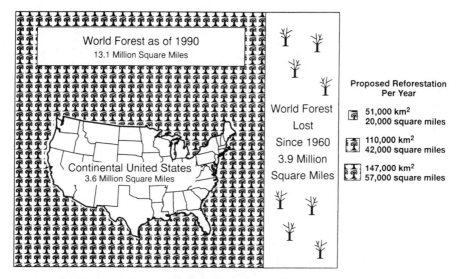

FIGURE 9-4. Worldwide potential of reforestation.

per year. However, before the power plants in the industrial countries reduce their CO_2 discharge, they first have to eliminate their increase of about 1.2% per year of 25% from 16 billion metric tons: $0.012 \times 0.25 \times 16$ billion = 48 million metric tons per year.

As we have seen in Chapter 8, a reduction of 5.85 million short tons per year (5.3 million metric tons per year) of CO_2 discharge from a single 750 MW lignite-fired power plant can be achieved by repowering it with natural gas-fired advanced gas turbines. Reducing the CO_2 discharge by $17 + 48 = 65$ million metric tons would only require the repowering of $65/5.3 = 12.3$ such power plants annually in the industrial world. Since building combined-cycle power plants and retiring old coal-fired power plants has a similar effect, it seems that the goal to stop the worldwide increase of CO_2 discharge from power plants can be achieved.

A major reduction of the carbon content in the atmosphere and therefore the greenhouse effect of CO_2 can be reached by reforestation of the already lost forest. As illustrated in Figure 9-4, a total of 3.9 million square miles of forest was lost between 1960 and 1990, which is more than the area of the continental United States. Even if the CO_2 discharge from fossil fuels would increase yearly by 100 million metric tons, or 1.6% of the 1990 level, a rise of the carbon content in the atmosphere

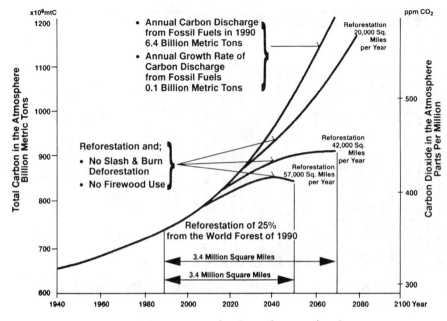

FIGURE 9-5. Atmospheric carbon reduction.

could be stopped. If 57,000 square miles of reforestation were provided each year, a reduction in atmospheric carbon would occur in about 2040, as shown in Figure 9-5. In 2050, 25% of the 1990 world forest would have been reestablished. This would still be less than the 30% forest area, which was lost from 1960 to 1990 by slash and burn deforestation.

The evolution of the power plant technology toward more efficient fuel conversion to electric power, starting with 15% efficient coal-fired power plants 100 years ago, is a very important aspect of the power generation industry. Figure 9-6 reveals the net power plant efficiency trend of various power plant concepts burning different fuels.

Highly efficient *pulverized coal-fired power plant concepts* are available with a plant net efficiency of 45% when operating with steam temperatures below 600°C (1110°F). This technology is available today and steam turbine power plants with even 640°C (1180°F) have been in operation since the late 1950s. Most advanced pulverized coal-fired power plants could be developed to operate at steam temperatures in the 700°C (1290°F) range. Such advanced power plants could be built as double-reheat units and could reach a net efficiency in the 50% range. With such

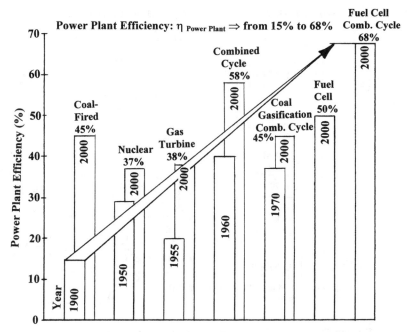

FIGURE 9-6. Present net power plant efficiency levels of various plant concepts burning different fuels (based on LHV).

coal-fired power plants, the CO_2 discharge could be reduced by 40% over presently operating coal-fired power plants.

Nuclear power plants provide the lowest power plant net efficiency of about 37%, utilizing light-water reactors as steam supply systems. The need for a higher efficiency was not a primary concern, unlike safety and reliability, of nuclear power plants because nuclear fuel has been available at relative low costs and there are no emissions. However, more advanced nuclear power plant concepts can be built. An example is the gas-cooled reactor plant shown in Figure 9-7. The two gas-cooled reactors feature an internal helium cycle. Those reactors could be built as relatively small, compact, and extremely safe 200 $MW_{Thermal}$ units and could be used in any number depending on the power plant size. Heat exchangers of each reactor transfer the heat energy to a nitrogen cycle for the gas turbine operating with 63 bar (900 psig) and 840°C (1550°F) inlet conditions. The closed nitrogen cycle has a pressure ratio of 3.4:1 and the gas turbine generates about 78 MW, with an exhaust temperature of about 595°C (1100°F) for the steam cycle. The steam turbine receives

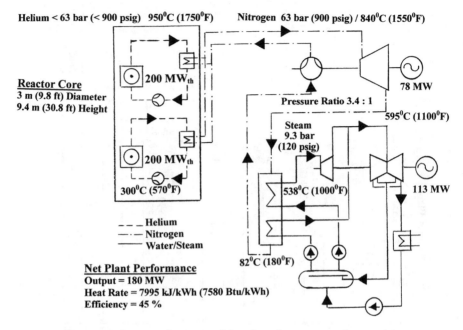

FIGURE 9-7. Nuclear combined-cycle power plant with helium-cooled reactor modules. (Source: Siemens AG)

main steam of 9.3 bar (120 psig) and 538°C (1000°F). The net output of a power plant with two reactor units would be about 180 MW with a net power plant efficiency of 45%.

Simple-cycle gas turbines are utilized to supply peaking power. Today, the efficiency of such plants can reach 38% for advanced heavy-duty gas turbines and 41% for the largest aero-engines. Efficiency for peaking turbines with less than 1000 full-load operation hours, in some applications of only 100 hours, should not be a major concern. However, improvement by increasing the firing temperature would be a possibility. This is especially the case if an increase in NO_X emission can be accepted for short peaking demands. Dry low-NO_X emission control concepts have been developed to achieve single-digit ppm NO_X values, but with increased combustion temperatures the achievable NO_X emission level also increases.

Combined-cycle gas/steam turbine plants have already achieved a power plant efficiency level of 58% and there is the potential to reach power plant efficiencies of 60% or more with the introduction of even more advanced technologies. More efficient fuel utilization can be achieved by utilizing any of these power plants as co-generation facilities.

Coal gasification combined-cycle (IGCC) power plants with present technology reach a net efficiency level of about 45%. There is a potential for further improvement by either the development of hot gas clean-up systems and/or by optimal integration of the three major plant sections (coal gasification, air separation, and power generation) and by advancing their performance.

IGCC power plant concepts cannot only be considered for co-generation, but also to supply marketable by-products. Figure 9-8 shows two examples of electrolysis/IGCC power plants with entrained flow gasification producing fertilizer in the form of urea or automobile fuel in the form of methanol or gasoline — all products well in demand.[76] Coal gas of an IGCC plant could be converted by means of the shift reaction $CO + H_2O \Rightarrow CO_2 + H_2$. After complete separation, hydrogen is burned in the gas turbine free of any CO_2 discharge, whereas the remaining divaricated carbon can be reused as a component for the synthesis of methanol. This alternative automobile fuel is obtained by the process of converting $CO_2 + 3H_2 \Rightarrow CH_3OH + H_2O$. If methanol cannot be easily marketed, the so-called MTG process can produce gasoline from the methanol.

Another potential by-product would be urea as fertilizer for the agricultural industry. Urea is formed by the following process $CO_2 + 2NH_3 \Rightarrow CO(NH_2)_2 + H_2O$. Since the demand for urea worldwide is large, and the storage and transportation of urea is not difficult, this might be an attractive co-product process for future IGCC plants. These processes need additional hydrogen that can be obtained from CO_2-free power generation sources, e.g., nuclear or hydro power stations during off-peak times. Such co-production plants could be built in the vicinity of nuclear or hydro power stations to receive the additional electricity for their electrolysis. The power plant efficiency of the 600 MW IGCC power plant is reduced by only a few percentage points. The gas turbine does not discharge any CO_2. The co-production of electric power and 1.3 million gallons of gasoline per day minimizes the overall CO_2 discharge, and oil as a raw product for gasoline could be substituted by coal as a long-term perspective.

Based on the proven recoverable fossil fuel reserves and their present use, it seems we will run out of fuel oil in about 45 years. As additional reserves not too much raw oil can be expected, but there is a large amount

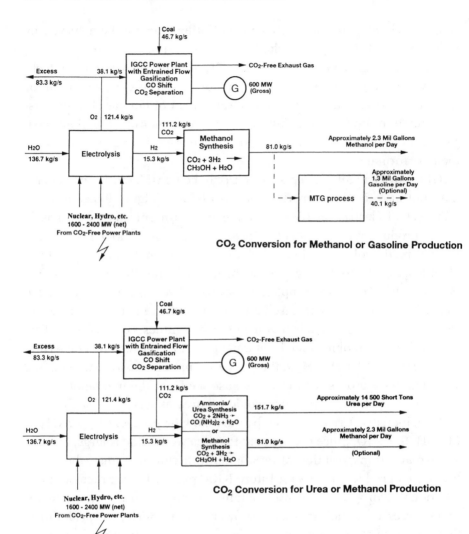

FIGURE 9-8. Electrolysis/IGCC power plants.

of oil shale, which can become the second largest fossil fuel supply after coal. To utilize these large reserves combined-cycle power plants and oil production facilities could be built, as illustrated in Figure 9-9. From 10,000 metric tons per day of raw oil shale about 130 MW electric power as the net output of the facility and an oil production of 4300 barrels per day could be provided. Such plants feature a hydropyrolysis section with a fluidized bed reactor, a product separation and heat recovery section,

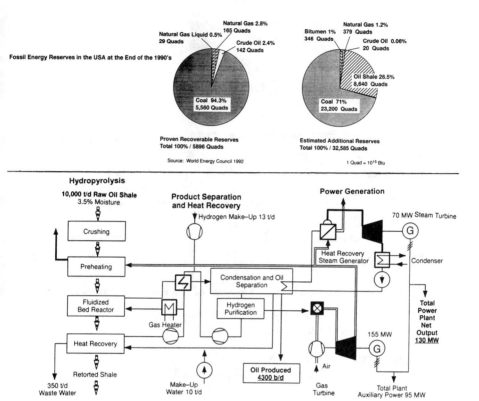

FIGURE 9-9. Combined-cycle power plant with oil production from the potentially large oil shale reserves.

and a power generation section with a 155 MW gas turbine and a 70 MW steam turbine. From the 95 MW auxiliary power a major portion is needed for the production of the hydrogen make-up of 13 metric tons per day.[1]

Fuel cells with their attractive efficiency around 50% are in the pilot plant stage for power generation. Presently, they are not quite as cost effective, but this could change in the near future — as soon as more efficient manufacturing processes are developed. Fuel cells because of their relatively low power density are not well suited for use in building large central power stations; however, they could become very attractive as distributed power generation units.

A major step toward even more efficient power generation plants can be achieved by combining a *fuel cell facility with a combined-cycle gas*

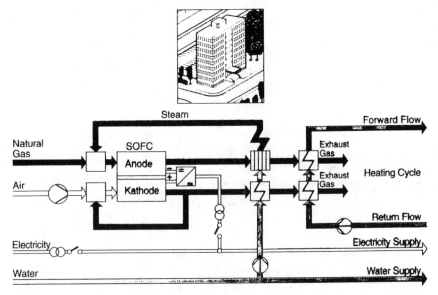

Plant Efficiency $\eta_{Electric} = 50\ \%$; $\eta_{Heat} = 33\ \%$
Fuel Utilization $\eta_{Plant} = 83\ \%$

FIGURE 9-10. Solid oxide fuel cell (SOFC) co-generation plant. (Source: Siemens AG)

turbine/steam turbine plant. As shown in Figure 8-10, such power plants with solid oxide fuel cells can achieve a power plant efficiency of 68%.

Another potential application of fuel cells would be an arrangement like the one shown in Figure 9-10, in which the solid oxide fuel cells provide electricity to a city building complex, hospitals, or public buildings. The SOFC discharge heat is used for the hot water and heating supply of the buildings. Such an arrangement could provide 83% fuel utilization with the electric power being generated at the SOFC efficiency level of 50%, plus 33% efficiency of the hot water and heating supply.

In the future such highly efficient plants when utilized for co-generation or co-production could be built as power sources that are widely deployed. This would provide a more distributed power supply system, which would help to reduce the need for additional power transmission from centralized power stations.

A potential of combining power generation is the combination of fossil and renewable energy sources. An example is illustrated in Figure 9-11, in which an 88 MW combined-cycle gas turbine/steam turbine power plant has been designed to receive low-pressure steam from a so-

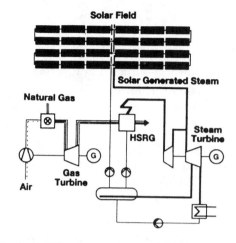
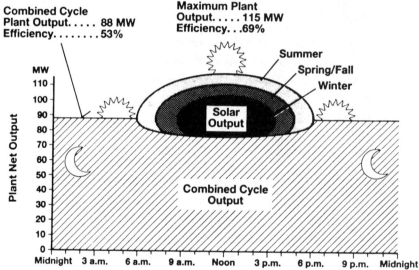

FIGURE 9-11. Combined-cycle power plant with solar LP steam.

lar field. At night the combined-cycle power plant generates the 88 MW at an efficiency level of 53%. During the day, when the power demand increases, the increased ambient temperature would result in less output at a lower efficiency. However, the demand for power increases considerably for air-conditioning use. With the solar LP steam power augmentation, the output and efficiency reaches its maximum of 115 MW and 69% in the summer at noon when electricity is most needed. A direct evaporation solar system has been developed in which steam is generated from water that is injected into a collector pipe system. The solar field of collector pipes is about 1000 m (3300 ft) long and 600 m (2000 ft) wide.[1]

There are certainly other power generation options with the available renewable fuel sources. Photovoltaics and wind power, for example, are already used for power generation. Even though these supplies are fast growing they can only cover a small portion of the total power supply. Such systems are ideally suited for power generation at remote locations where power transmission and distribution cannot be economically justified. They also could become a small part of a decentralized power generation system.

Presently, the activities in the power generation industry are concentrated in building mainly combined-cycle power plants and some simple-cycle and co-generation power plants. This trend is supported by the following influential factors:

- power generation industry deregulation
- trend toward distributed power supply
- low natural gas fuel costs
- low emissions and CO_2 discharge
- low specific plant costs
- short plant construction time
- compact plant design

Because the financing of such short-term projects is relatively easy, a large number of independent power suppliers entered the deregulated market of power generation by building such power plants.

Building coal-fired and nuclear power plants came practically to a standstill. However, more than half of all the electric power provided today is generated in relatively old coal-fired and nuclear power plants with an efficiency of less than 35%. Since not many large coal-fired and nuclear power plants have been built in the last 25 years, most of our electric power is generated in such power plants built before 1975.

The obvious shortcoming in power generation today is taken care of by building simple-cycle and combined-cycle power plants, which burn natural gas. Predictions have been made that distributed power generation with small, mini, and micro gas turbines, as well as fuel cells, will become a major electric power source of the future. The final solution of some predictions would be the fuel cell *black box*, which can be installed in each house to provide electricity, heat, air conditioning, and hot water. However, these power generation concepts require natural gas distribution to replace the present electricity transmission and distribution. Today, the technology of such a black box concept seems to be within reach, but the question is, is it cost effective and where is all the natural gas coming from.

Decentralized or distributed power generation with small plants is a very good approach, if these plants can be built as co-generation stations with a high fuel utilization. Building a large number of small gas turbine power plants or hybrid SOFC + GT power plants instead of one centralized full-size gas turbine power plant can hardly be justified when comparing the power density of these concepts. A 5 kW micro gas turbine-generator is about 0.3 m (1 ft) long and a full-size 170 MW gas turbine-generator has a length of 26 m (85 ft). The specific power per gas turbine-generator length in meters or feet can be calculated to be:

Micro Gas Turbine

5 kW/0.3048 m = 16.4 kW/m (5 kW/1 ft = 5 kW/ft)

Full-Size Gas Turbine

170,000 kW/26 m = 6540 kW/m (170,000 kW/85 ft = 2000 kW/ft)

To generate the 170 MW output of one full-size gas turbine with 5 kW micro gas turbines requires the installation of a total of 170,000 kW/5kW = 34,000 units. When lining up these 34,000 units, the line of micro gas turbines would be 10.4 km or 4.1 miles long. The notion that deregulation will lead to a competitive market with lower electricity costs cannot be supported. Experience shows that the cost for household electricity drastically increased after deregulation was introduced in Southern California.

The automobile industry is considering the application of fuel cells and hydrogen as fuel for buses, cars, and trucks. Electric cars are another available option. If one has visited Mexico City, the need for such a change is obvious. The power generation industry can support this trend by providing low-priced electricity at low-demand times. In addition, power plants could produce hydrogen by electrolysis when the demand for electricity is low. The power generation industry could also produce oil and gasoline as by-products from coal or oil shale. The combination of power generation and supply of high quality fuel can help prevent potential future shortages.

Nuclear power generation is needed because of the almost unlimited nuclear fuel supply. More efficient and even safer nuclear power plant concepts are available today, when compared to the technology of the presently operating nuclear plants, which were designed in the 1960s and 1970s.

Building only natural gas-fired power plants cannot continue much longer. The power generation industry must establish a long-term plan, which is based on long-term fuel resources. Natural gas will become very expensive and even scarce if the present trend of the power generation industry continues. The price of natural gas in the U.S. has about doubled within the last year. Drastic electric power price increases can be avoided by power generation systems with sufficient fuel flexibility.

Oil and natural gas should be wisely used by the power industry because the resources are limited and other users are more dependent on these limited resources. The challenges for the engineer in the power generation industry today are endless because a wide variety of advanced technologies are in the developmental stage. Most of today's electric power is generated in very old power generation stations, which must be rebuilt by adopting advanced technologies or they have to be replaced. The increasing demand for electric power, especially in developing countries, requires building facilities that meet the needs of each specific application and that use fuel that is locally available. Replacing wood burning for lighting, heat, and cooking with electricity from relatively simple and low-cost coal-fired power plants already provides a significant local improvement of the air quality.

The mission of the future global power generation industry should be to generate electricity for all people in the world. If the power generation industry sufficiently supports research, development, and engineering to advance all types of power generation options with any possible fuel source, this goal could be achieved. Power generation under various conditions in different countries requires building not necessarily the most efficient power generation facilities, but locally the most effective ones within the worldwide ecological restraints.

Time is the most significant factor for the power generation industry. Long-term global planning is absolutely necessary because the most advanced and most suitable power plants using different kind of fuels and fitting different applications cannot be developed and built overnight.

REFERENCES

[1] Riedle, K, Kuenstle, K., and Termuehlen, H.,"Power Generation and Its Effect on the Environment," *Proc. of Power-Gen International Conference*, Dallas, TX, 1993.

[2] Kraftwerk Union, "75 Jahre Turbinenfabrik Berlin," KWU X8-819, Berlin, Germany, 1979.

[3] Berliner Kraft-und Licht (Bewag)-Aktiengesellschaft, "Kraftwerk Moabit," Berlin, Germany, 1990.

[4] Bolter, J. R., "1994 Parsons Memorial Lecture: Sir Charles Parsons and electrical power generation—a turbine designer's perspective," Institution of Mechanical Engineers, meeting in London, UK, 1994, Proceedings of IMechE, Vol. 208, pp. 5–94.

[5] Stodola, A., *Die Dampfturbine*, Julius Springer Verlag, Berlin, Germany, 1910.

[6] Pink, H.-R., Ludewig, M. et al., *Steam Turbines and Turbinegenerators*, AEG, Berlin/Frankturt, Germany, 1963.

[7] Sass, F., Bouche, Ch., and Leitner, A., *Dubbels Taschenbuch fuer den Mashinenbau*, Springer Verlag, Berlin, Heidelberg, New York, 1966.

[8] Haas, H., Engelke, W., Ewarld, J., and Termuehlen, H.,"Turbines for Advanced Steam Conditions," *Proc. of American Power Conference*, Chicago, IL, 1982.

[9] Haas, H., Zimmermann, A., and Termuehlen, H., "Turbines for Advanced Steam Conditions: Operational Experience and Development," Electric Power Research Institute (EPRI) Conference on Improved Coal-Fired Power Plants, Palo Alto, CA, 1986, EPRI-Proceedings CS-5581.

[10] Campbell, C., Franck, C., and Spahr, J., "The Eddystone Superpressure Unit," *ASME Transactions*, **79,** pp. 1431–1446, 1957.

[11] Dierman, H. Kuhl, F., and Galloway, T., "Ravenswood 1000 MW Unit 3," *Proc. of American Power Conference*, Chicago, IL, 1963.

[12] Bergmann, D., Oeynhausen, H., and Termuehlen, H., "Large Steam Turbines for Advanced Power Plants," *Proc. of American Power Conference*, Chicago, Illinois, 1993.

[13] Oeynhausen, H., Drosdziok, A., Ulm, W., and Termuehlen, H., "Advanced 1000 MW Tandem-Compound Reheat Steam Turbine," *Proc. of American Power Conference*, Chicago, IL, 1996.

[14] Pollmann, E., and Termuehlen, H., "Turbine Rotor Vibrations Excited by Steam Forces (Steam Whirl)," ASME Winter Annual Meeting, Houston, TX, 1975, ASME Paper 75-WA/PWR-11.

[15] Pollmann, E., Schwerdtfeger, H., and Termuehlen, H., "Flow Excited Vibrations in High-Pressure Turbines (Steam Whirl)," ASME Winter Annual Meeting, Atlanta, GA, 1977, *ASME Journal of Engineering for Power*, April 1978, **100**.

[16] Neumann, K. Stannowski, G., and Termuehlen, H., "Thirty Years Experience with Integrally Shrouded Blades," *Proc. of ASME Joint Power Generation Conference*, Dallas, TX, 1989, ASME PWR-Vol. 7.

[17] Gloger, M. Neumann, K., and Termuehlen, H., "Design Criteria for Reliable Low-Pressure Blading," ASME International Joint Power Generation Conference, Portland, OR, 1986, ASME Paper 86-JPGC/PWR-42.

[18] Gloger, M. Neumann, K. Bergmann, D., and Termuehlen, H., "Advanced LP Turbine Blading: A Reliable and Highly Efficient Design," ASME International Joint Power Generation Conference, Atlanta, GA, 1992, ASME Power Division Proceedings.

[19] Termuehlen, H., "Variable-Pressure Operation and External Bypass Systems to Improve Power Plant Cycling Performance," ASME Joint Power Generation Conference, Charlotte, NC, 1979, ASME Paper 79-JPGC/PWR-9.

[20] Termuehlen, H., *Fossil-Fueled Power Plant Design and Operation—European Style*, Rocky Mountain Electrical League, Cheyenne, WY, 1982.

[21] Muehle, E-E., Keienburg, K-H., and Termuehlen, H., "Considerations to Achieve Reliable Long-Time Turbine Operation," ASME Joint Power Generation Conference, Miami, FL, 1987, ASME Paper 87-JPGC/PWR-55.

[22] Schleithoff, K., and Termuehlen, H., "Steam Purity in German Power Plants: Standards, Operational Data and Effects on Turbine Com-

ponents," ASME Joint Power Generation Conference, Dallas, TX, 1978, ASME Paper 78-JPGC/PWR-20.

[23] Engelke, W. Schleithoff, K., Jestrich, H-A, and Termuehlen, H., "Design, Operation and Inspection Considerations to Control Stress Corrosion of LP Turbine Disks," *Proc. of American Power Conference,* Chicago, IL, 1983.

[24] Schleithoff, K,. Neumann, K,. and Termuehlen, H., "Advanced Disk-Type LP Turbine Rotor Experience," Electric Power Research Institute (EPRI) Workshop Fossil Steam Turbine Disk Cracking, Charlotte, NC, 1990, EPRI Proceedings GS-7250.

[25] David, W., Roettger, G., Schleithoff, K,. Hamel, H., and Termuehlen, H., "Disk-Type LP Turbine Rotor Experience," ASME International Joint Power Generation Conference, Kansas City, MO, 1993, ASME Power Division Proceedings.

[26] Engelke, W., Boeer, J., Bergmann, D., and Termuehlen, H., "Turbine-Generators for 400 MW Coal-Fired Power Plants," ASME International Joint Power Generation Conference, San Diego, CA, 1991, ASME Power Division Proceedings.

[27] Alkire, G., Galland, H., and Termuehlen, H., "LP Turbine Rotor and Inner Casing Replacement," ASME International Joint Power Generation Conference, Philadelphia, PA, 1988, ASME PWR-Vol.3.

[28] Kehr, M., Emsperger, W., and Termuehlen, H., "Emission Reduction Program in the Unified Germany," *Proc. of American Power Conference,* Chicago, IL, 1992.

[29] Weschenfelder, K-D., Oeynhausen, H., Bergmann, D., Hosbein, P., and H.Termuehlen, "Turbine Steam Path Replacement at the Grafenrheinfeld Nuclear Power Station," *Proc. of American Power Conference,* Chicago, IL, 1994.

[30] Schleithoff, K., and Termuehlen, H., "Advancements in Nuclear Steam Turbine Design to Prevent Corrosion," *Proc. of American Nuclear Society Conference,* San Diego, CA, 1988.

[31] Lyle, F., Burghard, H., and Kolar, M., "Steam Turbine Disc Cracking Experience," EPRI Report NP- 2429-LD, Vol. 1, 1982.

[32] Jacobsen, G., Oeynhausen, H., and Termuehlen, H., "Advanced LP Turbine Installation at the 1300 MW Power Station Unterweser," *Proc. of American Power Conference,* Chicago, IL, 1991.

[33] Oeynhausen, H., Roettger, G., Meade, W., and Termuehlen, H., "Replacement of Nuclear LP Turbine Rotors and Inner Casings," *Proc. of American Power Conference*, Chicago, IL, 1997.

[34] Stambler, I., "DLE and Digital Controls Give New Life to Uprated 525 kW and 3000 kW Designs," *Gas Turbine World*, January-February 1997.

[35] Becker, B., Balling, L., and Termuehlen, H., "Advanced Siemens Gas Turbines and Their Applications," 12th Symposium on Industrial Applications of Gas Turbines, Banff, Canada, 1997, sponsored by Canadian Gas Association.

[36] Schenk, K., and Zoerner, W., "Advanced Steam Turbines with Heat Extraction Walsum and Herne Power Plants," ASME International Joint Power Generation Conference, Boston, MA, 1990, ASME PWR-Vol.11.

[37] Johnson, T., Becker, B., Seume, J., and Termuehlen, H., "First V 84.3A Gas Turbine Installation at Hawthorn Station," ASME Turbo-Expo, Orlando, FL, 1997, ASME Paper 97-GT-357.

[38] Friedrich, R., *Dokumente zur Erfindung der heutigen Gasturbine vor 118 Jahren*, VGB-Kraftwerkstechnik GmbH, VGB-B 100, Essen, Germany, 1991.

[39] Meher-Homji, C., "The Development of the Junkers Jumo 004B: The World's First Production Turbojet," ASME International Gas Turbine and Aeroengine Congress and Exhibition, Birmingham, UK, 1996, ASME Paper 96-GT-457.

[40] Way, S., "Comments on the Development of the Early Westinghouse Turbojets 1941-1946," ASME International Gas Turbine and Aeroengine Congress and Exposition, Cincinnati, OH, 1993, *ASME Transactions*, **116**, pp. 315–321.

[41] Brandt, D.E., "MS7001F Prototype Test Results," ASME Gas Turbine and Aeroengine Congress and Exposition, Toronto, Ontario, Canada, 1989, ASME Paper 89-GT-102.

[42] Scalzo, A., Bannister, R., De Corso, M., and Howard, G., "Evolution of Heavy-Duty Power Generation and Industrial Combustion Turbines in the United States," International Gas Turbine and Aeroengine Congress and Exposition, The Hague, Netherlands, 1994, ASME Paper 94-GT-488.

[43] Becker, B., Schulenberg, T., and Termuehlen, H., "The 3A-Series Gas Turbines with HBR Combusters," ASME Turbo Expo, Houston, TX, 1995, ASME Paper 95-GT-458.

[44] Becker, B., Huettenhofer, K., and Termuehlen, H., "Evolution of Gas Turbines from the .2 to the .3A Series Technology," Power-Gen, Latin American Power, Buenos Aires, Argentina, 1998.

[45] Schulenberg, T., "Key Technologies for Improving Gas Turbine Performance," *Siemens PowerJournal*, International Edition 2/95, Erlangen, Germany, August 1995.

[46] Becker, B., Huettenhofer, K., and Termuehlen, H., "Operating Experience and Upgrades of V 84.3A Gas Turbines," Power-Gen International, Orlando, FL, 1998.

[47] Becker, B., Balling, L., and Termuehlen, H., "Advanced Siemens V 84.3A and V 94.3A Gas Turbines," Power-Gen International, Dallas, TX, 1997.

[48] Becker, J., Kugler, J., Maghon, H., Schellhorn, L., and Termuehlen, H., "Gas Turbine Operating Performance and Considerations for Combined Cycle Conversion at Hay Road Power Station," *Proc. of American Power Conference*, Chicago, IL, 1990.

[49] Becker, B., Hoffmann, S., and Termuehlen, H., "Advanced V 84.3A Gas Turbines Operating Experience with Dry Low-NO_X Combustion," Power-Gen International, Orlando, FL, 1998.

[50] Maghon, H., Kreutzer, A., and Termuehlen, H., "The V 84 Gas Turbine Designed for Reliable Base Load and Peaking Duty," *Proc. of American Power Conference*, Chicago, IL, 1988.

[51] Schetter, B., Schabbehard, H., and Termuehlen, H., "Hybrid Burner Combustion of Fuel Oil in Premix Mode," *Proc. of Power-Gen Americas*, Orlando, FL, 1994.

[52] Maghon, H., Becker, B., Schulenberg, T., Kraemer, H., and Termuehlen, H., "The Advanced V 84.3A Gas Turbine," *Proc. of American Power Conference*, Chicago, IL, 1993.

[53] Joos, F., Brunner, B., Schulte-Werning, B., Syed, K., and Erogln, A., "Development of the Sequential Combustion System for the ABB GT 24/GT 26 Gas Turbine Family," ASME International Gas Turbine and

Aeroengine Congress and Exhibition, Birmingham, UK, 1996, ASME Paper 96-GT-315.

[54] Mayer A, and S. van der Linden, S., "GT 24/26 Advanced Cycle System Power Plant Progress for the New Millenium," ASME International Gas Turbine and Aeroengine Congress and Exhibition Indianapolis, IN, 1999, ASME Paper 99-GT-404.

[55] Corman, J., "'H' Gas Turbine Combined Cycle Technology and Development Status," ASME International Gas Turbine and Aeroengine Congress and Exhibition, Birmingham, UK, 1996, ASME Paper 96-GT-11.

[56] McQuiggan, G., "Designing for High Reliability and Availability in New Combustion Turbines," International Gas Turbine and Aeroengine Congress and Exhibition, Birmingham, UK, 1996, ASME Paper 96-GT-14.

[57] GE Marine and Industrial Engines, "LM6000 Gas Turbine," AE-3574 (11/96), Cincinnati, Ohio.

[58] Rodgers, A., "25-5 kWe Micro Turbine Design Aspects," ASME International Gas Turbine and Aeroengine Congress and Exhibition, Munich, Germany, 2000, ASME Paper 2000-GT-626.

[59] Boehm, H., "Fossil-Fired Power Plants," VGB Kraftwerkstechnik 74, Essen, Germany, 1994.

[60] Becker, B., and Termuehlen, H., "Evolution of Siemens Gas Turbine Technology," *Proc. of Power-Gen International,* Orlando, FL, 1996.

[61] Maghon, H., Brueckner, H., Bergmann, D., Kriesten, W., and Termuehlen, H., "Combined Cycle Power Plants for Load Cycling Duty," *Proc. of American Power Conference,* Chicago, IL, 1989.

[62] Engelke, W., Bergmann, D., and Termuehlen, H., "Steam Turbines for Combined Cycle Power Plants," ASME International Joint Power Generation Conference, Boston, MA, 1990, ASME PWR-Vol.11.

[63] Oeynhausen, H., Bergmann, D., Balling, L., and Termuehlen, H., "Steam Turbine Development for Advanced Combined Cycle Power Plants," ASME International Joint Power Generation Conference, Houston, TX, 1996.

[64] Ham, A., and Gloger, M., "High Back-Pressure Turbine Experience in 4000 MW Kendal Power Station," ASME International Joint Power Generation Conference, Boston, MA, 1990, ASME PWR-Vol. 11.

[65] Utter, R., Rohwer, H-O., Brueckner, H., and Termuehlen, H., "Combined Cycle Power Plant Genelba in Argentina," *Proc. of Power-Gen International*, Orlando, FL, 1996.

[66] Zabrecky, J., Bezugly, J., Brown, M., and Martin, H., "High Power Density, 60 Hz, Single Flow Steam Turbine with 42 inch Titanium Last Row Blade for Advanced Combined Cycle Applications," ASME International Joint Power Generation Conference, San Francisco, CA, 1999, ASME PWR- Vol. 34.

[67] Daledda, K., Rogers, T., Meyer, R., and Termuehlen, H., "Repowering of Steam Power Plants with Gas Turbines in the United States," *Proc. of Power-Gen Europe*, Koeln, Germany, 1994.

[68] Pfost, H., Rukes, B., and Termuehlen, H., "Repowering with Gas Turbines Utilizing Their Exhaust Energy for Feedwater Heating and/or Reheat Steam Generation," ASME International Joint Power Conference, Denver, CO, 1997.

[69] Termuehlen, H., "Repowering Steam Turbine Plants by Adding Gas Turbines," Electric Power Research Institute (EPRI) Workshop Repowering Technology Assessment, Ft Lauderdale, FL, 1993.

[70] Achenbach, J., Kraemer, H., and Termuehlen, H., "Repowering Existing Power Stations with Heavy- Duty Gas Turbines: An Economical Approach," *Proc. of Power-Gen International*, 1992.

[71] Haupt, H., Thiel, H.-J., and Termuehlen, H., "Repowering of the Peterhead Power Station Provides Operating Flexibility, Increased Efficiency and Emission Reduction," ASME International Joint Power Generation Conference, San Francisco, CA, 1999, ASME PWR-Vol. 34.

[72] Termuehlen, H., "Repowering: An Effective Way to Reduce CO_2 Discharge," ASME International Joint Power Generation Conference, Baltimore, MD, 1998.

[73] Brueckner, H., Bergmann, D., and Termuehlen, H., "Various Concepts for Topping Steam Turbine Plants with Gas Turbines," *Proc. of American Power Conference*, Chicago, IL, 1992.

[74] Termuehlen, H., "Combined Cycle Power Plant Applications for Reducing Fuel Consumption and Emissions," Power-Gen Latin American Conference on Turbomachinery, Cuernavaca, Morelo, Mexico, 1993.

[75] Kreutzer, A., Ganzer, W., and Termuehlen, H., "Gas and Coal-Fired Combined Cycle Plants *Proc. of American Power Conference*, Chicago, IL, 1986.

[76] Mueller, R., and Termuehlen, H., "The Future of Integrated Coal Gasification Combined Cycle Power Plants," *Proc. of Power-Gen*, Tampa, FL, 1991.

[77] Emsperger, W., Haupt, G., and Karg, J., "IGCC Technology Provides Clean and Efficient Power," *Proc. of Power-Gen Asia*, New Delhi, India, 1996.

[78] Becker, B., Schetter, B., and Termuehlen, H., "Low-Emission Combustion in Advanced Gas Turbines with Conventional and Coal-Derived Fuel," Electric Power Research Institute (EPRI) 10th Annual Conference on Gasification Power Plants, San Francisco, CA, 1991.

[79] Campanari, S., and Macchi, E., "Performance Prediction of Small-Scale Tri-Generation Plants Based on Integrated SOFC and Micro Turbine Systems," ASME International Gas Turbine and Aeroengine Congress and Exhibition, Munich, Germany, 2000, ASME Paper 2000-GT-318.

[80] Veyo, S., Shockling, L., Dederer, J., Gillet, J., and Lundberg, W., "Tubular Solid Oxide Fuel Cell/Gas Turbine Hybrid Cycle Power Systems–Status," ASME International Gas Turbine and Aeroengine Congress and Exhibition, Munich, Germany, 2000, ASME Paper 2000-GT-550.

[81] Lezue, A., Riedle, K., and Wittchow, W., *"Development Trends in Hard Coal-Fired Power Plants,"* Brennstoff-Waerme-Kraft **41,** pp.13–23, VDI Verlag, Duesseldorf, Germany, 1989.

ABOUT THE AUTHOR

Heinz Termuehlen was born in Berlin, Germany in 1936. He went to school and studied mechanical engineering in Berlin. After graduating in 1958, he began employment with Allgemeine Elektricitaets Gesellschaft (AEG), later Siemens Kraftwerke Union (KWU), in the Steam Turbine Application and Project Department. In 1970, Mr. Termuehlen joined Allis-Chalmers Power Systems in Milwaukee, Wisconsin. In 1979 he became manager of application engineering with Utility Power Corporation in Bradenton, Florida. In 1994 he was promoted to the position of chief engineer and director of product planning for Siemens Power Corporation in Milwaukee, Wisconsin.

Mr. Termuehlen has authored 98 ASME, ANS, APC, Power-Gen, and IEEE papers. In 1980 and 1988, he was honored with the ASME Prime Movers Award for the best publication of the year. Mr. Termuehlen has been a member of the ASME Power Division's Committee on Steam Turbines, and the International Representatives Committee, as well as the Industrial Committee of the American Power Conference. In 1997, he formed the Combined-Cycle Committee of ASME Power Division and is presently chairperson.

In 1988, Mr. Termuehlen was awarded the ASME Life Fellowship. He has received various patents, among them two recent U.S. patents on repowering concepts of steam turbine plants with gas turbines.

Mr. Termuehlen has been active in the power generation industry worldwide for 41 years. Last year he retired from Siemens and wrote this book to ensure that the experience gained with "100 Years of Power Plant Development" would contribute to shaping the power generation industry of the future.

INDEX

A

Absorption chillers, 167
Adiabatic expansion, 20
Aero-engines, 86
Aeroderivative gas turbine co-generation plants, 67, 69
Aeroderivative gas turbines, peaking power plants with, 86
Aeroderivative mid-size gas turbines, 105–106
Aeroderivative mini gas turbines, 107–108
Aeroderivative small gas turbines, 107, 108
Air cooling, open-loop, 103
Air intake, 42
Alloy steels, 21
Annulus area, 118
Atmospheric carbon reduction, 178
Austenitic forgings and castings, 28
Automobile fuel, 181
Auxiliary efficiency, 20
Availability, fuel, 174

B

Backpressure of steam turbines, 22
Backpressure steam turbines, 68–69
Barrel-type outer casings, 28
Bearing designs, 36
Black box, fuel cell, 186–187
Blade coating, 92–93
Blade cracking, 36–37
Blast furnace gas, 85
Boiler efficiency, 20, 33, 76

Booster compressor, 157
Brayton cycle efficiency, 89
Brayton process, 88–89
    combined with Rankine process, 115–116
Burners, premix, 95
Bypass systems, combination of shut-off valves and, 141

C

C (Celsius) temperature, 15
Campbell diagrams, 37, 38
Carbon dioxide discharge, 48–49, 175–177
    reduction of, 142–143, 151–153, 176–178
Carbon reduction, atmospheric, 178
Carnot cycle, 12–18
Carnot cycle efficiency, 14–16
Carnot efficiency, combined-cycle power plants, 16
Cars, electric, 187
Catalytic converters, 47–48
Cavities, formation of, 40
Celsius (C) temperature, 15
Chernobyl accident, 63
Chillers, absorption, 167
Chilling systems, 98
China, 175
CHP (combined heat and power), 67
Clausius-Rankine process, 16
Closed-loop steam cooling, 103
Co-generation, 67
Co-generation heat, total, 73

Co-generation plants, 6, 67–81
   aeroderivative gas turbine, 67, 69
   combined-cycle, 68
      examples of, 80–81
      proper evaluation of, 74–75
   solid oxide fuel cell, 184
Coal, heat value of, 33
Coal combustion, pulverized, 25
Coal-fired power plant concepts, pulverized, 178–179
Coal-fired steam plant, parallel powered, with natural gas-fired gas turbine, 150–151
Coal gas, 181
Coal gasification plant, early, 157, 158
Combined-cycle co-generation plants, 68
   examples of, 80–81
Combined-cycle efficiency, 123
Combined-cycle power plants, 5, 113–130
   Carnot efficiency, 16
   efficiency, 180–181
   fuel cell, *see* Fuel cell combined-cycle power plants
   fuel cell facility with, 183–184
   increased output of, by power augmentation, 129–130
   major concepts, 114–115
   mass flow of, 118
   optimal, 147
   output of, 7
   performance improvement of, 114
   present trend towards, 173
   with pressurized fluidized bed combustion systems, 157
   with solar LP steam, 184–186
   steam cycles of, 117
   steam turbines for, 127
Combined heat and power (CHP), 67

Combustion, pulverized coal, 25
Combustion dust removal, 47
Combustion noise, 96
Combustion systems, pressurized fluidized bed, combined-cycle power plants with, 157
Combustion temperature, 101
Compressor, booster, 157
Compressor inlet temperature, 96–97
Computerized start-up systems, 39
Condensate polishing systems, 42
Condensate pump, 11
Condenser design, 42
Condensing steam turbines with extractions, 69
Cool Water project in California, 158
Cooling techniques, 93
Cooling towers, 22
Corrosion products, 45
Cracking, stress corrosion, 42–44
Creep, 40–41

D

Decentralized power generation, 187
Digital controls, 39
Discharge
   carbon dioxide, 48–49
   pollutant, 47
Discharge products, 166
Disk-type rotor design modifications, 60, 62
Distributed power generation, 187
District heating plant, heat balance diagram of, 79
Drain slots, 46
Drain systems, 22
Droplet erosion, 45–46
Drum-type boilers, 28
Dual-pressure cycle, 116

Dual-pressure reheat cycle, 138
Dust removal, 47

E

EB-PVD (electron beam physical vapor deposition) process, 93, 94
Economizer, 11
Edison, Thomas, ii, 1
Efficiency, 90
  boiler, 20, 33, 76
  Brayton cycle, 89
  Carnot cycle, 14–16
  combined-cycle, 123
  combined-cycle power plants, 180–181
  electrical net, 169
  electrical power, 73
  integrated coal gasification combined-cycle power plants, 163, 181
  nuclear power plants, 179
  power plant overall, 33
  present net power plant, 179
  reheat steam turbine, 146
  repowering concept, 138
  simple-cycle, of gas turbines, 107
  thermal, of steam turbines, 20
  total power plant, 146–147
Electric cars, 187
Electric power, 1
Electric power consumption, imbalance between world population and, 176
Electric power generation, 173
Electric Power Research Institute (EPRI), 59
Electrical net efficiency, 169
Electrical power efficiency, 73
Electrolysis/IGCC power plants, 182

Electron beam physical vapor deposition (EB-PVD) process, 93, 94
Electrostatic precipitators, 47
Enthalpy drop, 19
Enthalpy raise, 32
Entrained flow gasifiers, 159, 161
Environmental effect of power plants, 2
EPRI (Electric Power Research Institute), 59
Erosion, 45–46
Erosion-corrosion attack, 57–59
Erosion-corrosion rate, pH level and, 58
Exhaust loss curves, steam turbine, 119
Exhaust temperature of gas turbines, 146
Expansion lines, 21
  non-reheat steam turbine, 70–71
Extractions, condensing steam turbines with, 69

F

Failure modes, 36–46
  creep, 40–41
  erosion, 45–46
  high-cycle fatigue, 36–37
  low-cycle fatigue, 37–40
  rotor instability, 36
  stress corrosion, 41–44
Fast breeder reactor designs, 63
Feedwater heater repowering, 133, 145–153
Feedwater heating, 24–25, 117
Feedwater preheating, 27
Feedwater pump, 11
Feedwater system design, 31
Fertilizer, 181
Film cooling, 93, 94

Fire turbine, 85
Firing temperature, 101
   increasing, 180
Fixed bed gasifiers, 158–159, 161
Flame hardening, 46
Flashbacks, 96
Fluidized bed combustion systems, pressurized, combined-cycle power plants with, 157
Fluidized bed gasifiers, 159, 161
Forth Banks, 11
Fossil energy reserves, worldwide recoverable, 173–175
Fossil steam turbine power plants, 6–7, 11–49
   early, 13
   first, 11
   for highest steam conditions, 29
   machine house arrangements, 13
   repowering, with gas turbines, 133–153
Fracture toughness, 41
FU (fuel utilization), term, 74
Fuel availability, 174
Fuel burner, triple, for integrated coal gasification combined-cycle (IGCC) power plants, 165
Fuel cell black box, 186–187
Fuel cell combined-cycle power plants, 166–170
Fuel cell facility with combined-cycle power plants, 183–184
Fuel cells, 183
Fuel consumption, difference in, 134
Fuel flexibility, 142
Fuel shortages, 187
Fuel sources, predicted, 173, 174
Fuel utilization (FU), term, 74
Fuels
   automobile, 181

   gas turbine, heat values of, 164
   low-Btu, 164–165
Full-arc admission, 36
Full-size gas turbines, 187
Fully fired repowering, 135

G

Gas
   blast furnace, 85
   coal, 181
   natural, *see* Natural gas *entries*
Gas-cooled nuclear power plants, 179–180
Gas/steam turbine arrangement, single-shaft, 128
Gas-to-steam turbine ratio, 137
Gas turbine/HRSG/steam turbine arrangement, 147
Gas turbine/HRSG unit, 136, 137
Gas turbine power plants, 7, 85–110
   early, 85
   unit size of, 7
Gas turbine temperatures, 100
Gas turbine to steam turbine output ratio, 148–149
Gas turbines, 5, 67
   advanced, 86, 88, 90–92
   advanced heavy-duty full-size, with sequential combustion, 99
   aeroderivative, peaking power plants with, 86
   aeroderivative mid-size, 105–106
   aeroderivative mini, 107–108
   aeroderivative small, 107, 108
   exhaust temperature of, 146
   full-size, 187
   micro, *see* Micro gas turbines
   natural gas-fired, parallel powered coal-fired steam plant with, 150–151

repowering fossil steam turbine
    power plants with, 133–153
repowering projects with, 153
simple-cycle, 180
simple-cycle efficiency of, 107
size definitions, 104
smaller, 104
term, 89
Gasification suppliers, 158
Gasifiers, three basic types of,
    158–161
Gasoline, 181
Generator efficiency, 20
Global, *see also* Worldwide *entries*
Global planning, long-term, 188
Global power generation industry, 1
Greenhouse effect, 175–176
GT, *see* Micro gas turbines

H
Half-speed turbines, 53
Heat
  combined, and power (CHP), 67
  supplied, 76
  total co-generation, 73
Heat balance diagram, 31–33
  of district heating plant, 79
  of nuclear power plant, 56
Heat energy, utilized, 73
Heat rate (HR), term, 74
Heat rate calculation, 32–33
Heat recovery steam generators, *see*
    HRSG *entries*
Heat values of gas turbine fuels, 164
Heating, feedwater, 24–25, 117
Heating plant, district, heat balance
    diagram of, 79
HHV, *see* High-heat value
High-cycle fatigue, 36–37
High-heat value (HHV)
  of coal, 33

  of natural gas, 90
High pressure (HP) turbine, 23–24
High-temperature gas-cooled reactor
    designs, 63
High tensile stressing, 43–44
Hot windbox repowering, 133,
    144–145
HP (high pressure) turbine, 23–24
HR (heat rate), term, 74
HRSG repowering, 133, 136–144
  optimal, 147
HRSGs (heat recovery steam
    generators), 68, 80–81
Hydropower generation, 173
Hydropyrolysis section, 182

I
IGCC, *see* Integrated coal
    gasification combined-cycle
    power plants
Impulse-type blade profiles, 34, 35
India, 175
Industrial turbine design, 77
Industrial turbines, 76
Inlet temperature
  compressor, 96–97
  ISO, 101
Integrally shrouded blades, 34, 36
Integrated coal gasification
    combined-cycle (IGCC) power
    plants, 157–166
  designing, 161–162
  efficiency, 163, 181
  flowchart of, 162–163
  integration of gasifiers into, 160
  triple fuel burner for, 165
Intermediate pressure (IP) turbine, 24
ISO inlet temperature, 101

J
Jet engines, 86

K

Kelvin (K) temperature, 15

L

LHV, *see* Low-heat value
Light-water reactors, 53
Long-term global planning, 188
Low-Btu fuel, 164–165
Low-cycle fatigue, 37–40
Low-heat value (LHV)
  of coal, 33
  of natural gas, 90
Low pressure (LP) turbine, 23–24
LP steam bypass station, 124–125
LP (low pressure) turbine, 23–24

M

Mass flow conditions, 139–140
Methanol, 181
Micro gas turbines (GT), 167, 187
  with recuperators, 109
Micro steam turbines, 110
Microcracks, formation of, 40
Missile analyses, 59
Moabit power station, ii, 1
Moisture content, 21
Moisture removal, internal, 54
Moisture separator/reheaters
  (MSRs), 54, 56–58
Mollier diagram, 19
Monoblock rotors, 60
MSRs (moisture
  separator/reheaters), 54,
  56–58
MTG process, 181
Multiple extraction turbine, reheat
  steam turbine as, 78

N

Natural gas, 188

Natural gas-fired gas turbine,
  parallel powered coal-fired
  steam plant with, 150–151
Natural gas fuel, switch to, 144
Nitrogen cycle, 179–180
Nitrogen dioxide emissions,
  reduction of, 47
Non-reheat steam turbine expansion
  lines, 70–71
Non-reheat steam turbines, reheat
  steam turbines converted
  into, 136
NRC (Nuclear Regulation
  Commission), 59
NSSS (nuclear steam supply
  systems), 60
Nuclear power generation, 188
Nuclear power plants, 5, 53–63
  efficiency, 179
  evolution process for, 7
  future, 173
  gas-cooled, 179–180
  heat balance diagram of, 56
  inherently safe, 63
  political issues and, 61–63
  reheat versus non-reheat, 54–56
Nuclear reactors, 2, 53
Nuclear Regulation Commission
  (NRC), 59
Nuclear steam supply systems
  (NSSS), 60

O

Oil, 188
Oil shale, 182
Once-through boilers, 28
Open-loop air cooling, 103
Output ratio, gas turbine to steam
  turbine, 148–149

P

Parallel powered coal-fired steam plant with natural gas-fired gas turbine, 150–151
Parallel repowering, 133, 145–153
Parsons, Charles, 11
Partial-arc admission, 36
Peaking power plants with aeroderivative gas turbines, 86
PEMFC (polymer electrolyte membrane fuel cell), 166
pH level, erosion-corrosion rate and, 58
Photovoltaics, 186
Planning, long-term global, 188
Political issues, 1
   nuclear power plants and, 61–63
Pollutant discharge, 47
Polymer electrolyte membrane fuel cell (PEMFC), 166
Population, world, *see* World population
Post-repowering pressure levels, 140
Power augmentation, increased output of combined-cycle power plants by, 129–130
Power generation
   decentralized, 187
   distributed, 187
   electric, 173
   future, 173–188
   nuclear, 188
   shortcoming in, 186
Power generation industry, global, 1
Power plant concepts, 1–2
   advances in, 5
Power plant output after repowering, 143
Power plant overall efficiency, 33
Power plants, 1
   co-generation, *see* Co-generation plants
   combined-cycle, *see* Combined-cycle power plants
   electrolysis/IGCC, 182
   environmental effect of, 2
   fossil steam turbine, *see* Fossil steam turbine power plants
   fuel cell combined-cycle, *see* Fuel cell combined-cycle power plants
   future, 173
   gas turbine, *see* Gas turbine power plants
   historical development of, 7
   integrated coal gasification combined-cycle, *see* Integrated coal gasification combined-cycle power plants
   nuclear, *see* Nuclear power plants
   present net efficiency, 179
   pulverized coal-fired, 178–179
   total, efficiency of, 146–147
   unit rating growth, 6, 7
Precipitators, electrostatic, 47
Precision casting technique, 92–93
Predicted fuel sources, 173, 174
Preheating, feedwater, 27
Premix burners, 95
Premix flame combustion noise, 96
Pressure, supercritical, 28
Pressure levels, post-repowering, 140
Pressurized fluidized bed combustion systems, combined-cycle power plants with, 157
Pulverized coal combustion, 25
Pulverized coal-fired power plant concepts, 178–179

Pulverized coal plant, 166
Pyrolysis, 158

R
R (Rankine) temperature, 15
Radial flow path design, 28
Rankine cycle of single reheat turbine, 24–25, 26
Rankine process, combined with Brayton process, 115–116
Rankine steam cycle, 16–18
Rankine (R) temperature, 15
Rathenau, Emil, ii, 1
Ravenswood power station, 29–30, 31
Reaction-type blade profiles, 34, 35
Recoverable fossil energy reserves, worldwide, 173–175
Recuperators, micro gas turbines with, 109
References, 189–196
Reforestation, worldwide potential of, 177–178
Refueling outage, 61
Reheat cycle
  dual-pressure, 138
  triple-pressure, 116, 122
Reheat steam turbines, 24–27
  converted into non-reheat steam turbines, 136
  efficiency, 146
  as multiple extraction turbine, 78
  two-casing single-flow, 128–129
Repowering
  feedwater heater, 133, 145–153
  fossil steam turbine power plants with gas turbines, 133–153
  fully fired, 135
  hot windbox, 133, 144–145
  HRSG, *see* HRSG repowering parallel, 133, 145–153
  power plant output after, 143
  three concepts of, 153
  topping, 133
Repowering application range, potential, 135
Repowering applications, 133–134
Repowering concept, efficiency, 138
Repowering projects with gas turbines, 153
Reserves, worldwide recoverable fossil energy, 173–175
Retrofitting, 40
Rotor design modifications, disk-type, 60, 62
Rotor inspection intervals, 59
Rotor instability, 36

S
Saturation line, 21
Scrubbers, 47, 48
Sequential combustion, advanced heavy-duty full-size gas turbines with, 99
Shale oil, 182
Shortages, fuel, 187
Shrunk-on disk-type rotors, 60
Shut-off valves, combination of bypass systems and, 141
Simple-cycle efficiency of gas turbines, 107
Simple-cycle gas turbines, 180
Single-crystal blading, 92–93
Single reheat turbine, Rankine cycle of, 24–25, 26
Single-shaft gas/steam turbine arrangement, 128
Slag, 166
SOFC (solid oxide fuel cell), 166–167
Solar LP steam, combined-cycle power plants with, 184–186

Solid oxide fuel cell (SOFC), 166–167
Solid oxide fuel cell (SOFC) cogeneration plant, 184
Solid particle erosion, 45–46
Start-up systems, computerized, 39
Steam cooling, 101–103
　closed-loop, 103
Steam cycles of combined-cycle power plants, 117
Steam injection, 96
Steam purity, 60
Steam turbine-generators, 11
Steam turbine output ratio, gas turbine to, 148–149
Steam turbine power plants, fossil, see Fossil steam turbine power plants
Steam turbine speed, 126
Steam turbines, 5
　backpressure, 68–69
　backpressure of, 22
　for combined-cycle power plants, 127
　condensing, with extractions, 69
　early design, 18
　exhaust loss curves, 119
　mass flow of, 118
　micro, 110
　non-reheat, reheat steam turbines converted into, 136
　reheat, see Reheat steam turbines
　tandem-compound, 35, 37
　thermal efficiency of, 20
Steam whirl, 36
Steels, alloy, 21
Stolze, Franz, 85, 87
Stress corrosion, 41–44
Stress corrosion cracking, 42–44
Stressing
　high tensile, 43–44
　thermal, 37, 39
Sulfur dioxide emissions, 47
Super-clean materials, 41
Supercritical pressure, 28
Superheating, 15
Supplied heat, 76

T

Tandem-compound steam turbines, 35, 37
Temperature
　Celsius (C), 15
　combustion, 101
　compressor inlet, 96–97
　coolant, 120, 121
　exhaust, of gas turbines, 146
　firing, see Firing temperature
　gas turbine, 100
　inlet, see Inlet temperature
　Kelvin (K), 15
　Rankine (R), 15
Temperature definitions, 100–101
Tensile stressing, high, 43–44
Termuehlen, Heinz, v, 197
Thermal efficiency of steam turbines, 20
Thermal stressing, 37, 39
Topping repowering, 133
Triple fuel burner for integrated coal gasification combined-cycle (IGCC) power plants, 165
Triple-pressure reheat cycle, 116, 122
Turbine component improvements, 34
Turbine enthalpy drop, 20
Turbine exhaust steam, 17
Two-casing single-flow reheat steam turbine, 128–129

U

Urea, 181
Utilized heat energy, 73

V

Vacuum plasma spray (VPS) coating process, 92–93
Volumetric flow, 24
VPS (vacuum plasma spray) coating process, 92–93

W

Water chemistry, 42
Water injection, 96

Welder rotors, 60
Wind power, 186
World population, 174
   imbalance between, and electric power consumption, 176
Worldwide, *see also* Global *entries*
Worldwide potential of reforestation, 177–178
Worldwide recoverable fossil energy reserves, 173–175

Y

Yield strength, 41

Cover Photo: VEAG, Schwarze Pumpe Power Station. Two lignite-fueled 800 MW supercritical units without smokestacks.

Main Steam Conditions: 253 bar (3670 psia), 544°C (1011°F)
Reheat Steam Conditions: 52 bar (754 psia), 562°C (1044°F)
Power Plant Electric Efficiency: $\eta_{Electric} \cong 41\%$
Power Plant Fuel Utilization When Supplying Process Steam and District Heating: $\eta_{FU} \cong 55\%$

Photograph courtesy of Siemens AG